# REVOLUTIONIZING CUSTOMER EXPERIENCE

★ ★ ★ ★ ★

TOP STRATEGIES FROM CX EXPERTS

GINA GIBSON & CHAD HENDREN

Library of Congress Control Number: 2024900321

# CONTENTS

# Dedication

Hello, my name is David Sudbey, and I am honored to have been asked to write the foreword for Gina and Chad's new book.

In the rapidly evolving landscape of today's business world, the mantra "customer is king" has never rung truer. As organizations navigate the complexities of a hyper-connected and customer-centric era, the ability to deliver exceptional experiences has become the hallmark of success. It is against this backdrop that Gina Gibson and Chad Hendren, two luminaries in the field of Customer Experience (CX), present a groundbreaking playbook in "Revolutionizing Customer Experience."

I have known Gina and Chad for more than a decade and during that time, I have been impressed by their passion to make an impact in companies and improve the way they not just think of but to also deliver CX. As someone who has worked in the CX field for more than 30 years, I can attest to the importance of and the valuable insights that the authors provide in this book. I am confident that readers will find this book to be a valuable resource, and I am excited to play a small part in introducing it to you.

Gina and Chad bring to bear a wealth of experience, distilled from years spent at the forefront of driving transformative change in global enterprises. This book is more than just a compendium of best practices; it is a strategic guide for organizations looking to metamorphose into truly customer-obsessed entities.

The narrative unfolds with a stark diagnosis of the stumbling blocks that impede the realization of exceptional customer experiences. Drawing on real-world case studies, the authors dissect the blind spots, cultural misalignments, and communication breakdowns that conspire to dilute the impact of even the most well-intentioned CX initiatives.

At the heart of their discourse are seven key impediments, each meticulously examined to reveal the underlying challenges organizations face. From the dominance of Total Cost of Ownership (TCO) and Return on Investment (ROI) to the perils of over-reliance on technology, Gibson and Hendren lay bare the obstacles that thwart CX excellence.

One of the central tenets of the book is the authors' call to recalibrate metrics to align with customer outcomes. The notion that "What You Measure is What You Get" (WYMIWYG) serves as a guiding principle, cautioning against the pitfalls of measuring the wrong metrics or doing so at incorrect frequencies. The authors offer tangible solutions, urging organizations to evolve their metrics, embrace Objective and Key Results (OKRs), and focus on metrics that truly resonate with customers.

The book doesn't merely stop at identifying challenges; it meticulously guides readers through the intricate process of cultural transformation. From securing leadership commitment to leveraging immersive business simulations, the authors present a blueprint for embedding CX excellence into the very fabric of organizational culture.

Perhaps most importantly, Gina and Chad emphasize that the pursuit of CX excellence is not a one-time endeavor but an ongoing cultural realignment. They underscore the need for continuous improvement, aligning organizational focus with the ever-evolving expectations of the modern customer.

"Revolutionizing Customer Experience" is more than a book; it is a clarion call for organizations to reinvent themselves in the crucible of customer-centricity. Through compelling storytelling and research-backed recommendations, the authors provide a compass for leaders serious about not only meeting but exceeding customer expectations.

This guidebook is a must-read for those ready and have the courage to embark on the journey of massively improving customer experiences and the sustained competitive advantage they bring.

David A Sudbey
Advisor, former President Cogito and Chief Customer Officer Genesys

# ACKNOWLEDGMENTS

Writing this book has been an incredible journey, one that exceeded our expectations and filled us with immense gratitude. We owe a debt of thanks to the people who made it all possible.

First and foremost, our families for their unwavering encouragement and understanding making this ambitious undertaking achievable.

We extend our heartfelt appreciation to Amy for hosting our writing sessions and listening to our never ending brainstorming!

Lydia, your creative genius added a touch of magic to our work, and for that, we are truly grateful.

To our friends —Ty, Tracy, Michelle, Shellie, Alex and Chris you have been our pillars of strength. Your unwavering support, encouragement, and, above all, your friendship, mean the world to us.

Last but not least, we are indebted to our customers. The experiences we've gained from working with you have been invaluable and have enriched our journey beyond measure. Thank you for being a part of our story.

# Prologue

Our industry suffers from the magic bullet mentality. We do "the" project, then think everything should be solved with new software or technology. But the greatest strides of improvement don't come from a new implementation. Continuous improvement comes from a shift in culture and a unified focus on customer experience (CX). And that's where this book comes in. It's time to follow the evolutionary CX journey of our day.

For in the vast landscape of the business world, where competition is fierce and customer expectations continue to evolve, a new chapter is unfolding. It is a story of transformation, innovation and the pursuit of exceptional customer experiences. This pursuit requires going on a remarkable journey, so you need to first prepare for what lies ahead.

First, you encounter Chad Hendren and Gina Gibson, two visionary experts in the realm of CX. Their paths have converged, their expertise intertwined and their shared passion for elevating the customer journey became the driving force behind their collaboration. Through their insights, you will learn that "exceptional customer experience" is not merely a buzzword or a fleeting trend but a critical differentiator that can shape the destiny of businesses.

Mirroring real-life scenarios, we have to ask ourselves: How did the customer become secondary, and what has happened in the past that landed us where we are today? Take this scenario for example:

> Tensions escalated. The time for business simulation Round II came to an abrupt close.
>
> "Ugh! There is too much work here, and we can't possibly get it all done," the engineering lead lamented.
>
> "Well, I'm not getting the feature that will increase my Customer Satisfaction (CSAT) this iteration. We are going to lose revenue again," the business owner retorted.
>
> Time was up, and the reality was evident: each department worked on what *they* thought was a priority. As the QA person began to review the committed deliverables for the sprint, it became apparent there wasn't one feature completed in its entirety; as a team, they had missed their targets for the sprint because each group focused on their own mission instead of the CEO's customer goals.

Everyone is tired of enduring poor customer experiences. While a lack of technology contributed to abysmal customer experiences in the past, that is not the case today; artificial intelligence/machine learning (AI/ML) has removed any technical barriers. However, all the technology in the world will not fix your CX problems; committing to putting your customer first, as the nucleus of your business will.

The contact center domain and your customer are both constantly evolving. As we help you learn a new way to deliver and measure your CX success, we will also introduce ways to help you stay relevant to your customers. Some of the information may not be brand-new concepts. What's new is what you are going to do with the information. Defining the CX for your enterprise, as I'm sure many of you have already figured out, is a never-ending journey. This book isn't about picking it up, reading it once, getting all the answers and letting it collect dust.

## What This Book Is and Isn't

If you think this book is going to tell you how to lower your Average Handle Time (AHT), increase your first-call resolution or any of the other standard metrics that have been measured in the contact center arena for so long, we are ecstatic to say, "This isn't that kind of book."

We aren't here to tell you which tools to use and what methodology or technology is best suited to your organization. The fact of the matter is, it doesn't matter. If Waterfall, Scrum, Agile, DevOps, ITIL or SAFe work for your organization — embrace it! The only thing that matters is your cultural investment and commitment to the continuous improvement process. The latest technology is not going to fix your customer experience issues. If you don't have a CX strategy revolving around your customer, you are making the journey to success much more difficult than it needs to be. Screw everything you think you must do because of convention — you won't succeed that way, and neither will your business.

This book is meant to be provocative — to open your mind to the possibility of departing from conventional contact center metrics.

## What You Will Get Out of This

In the pages that follow, you will witness the emergence of a new era, where organizations are awakening to the strategic imperative of placing the customer at the heart of their operations.

This book challenges you to measure the business value of what you want to achieve without focusing on technology and instead, focusing on true metrics of meaning. We will explore releasing the shackles of old-school metrics and opening the door to exploring the possibilities of the new end state, what it looks like and the steps you can take to get there. This isn't necessarily an easy concept to grasp. Even as we compiled our thoughts for the book, we constantly found ourselves falling into the pitfalls of being led by technology and business needs — it's ingrained in us all.

We will address where most companies are today and offer ways to think differently in the future concerning how to address the CX in your company. Foundationally, this includes aligning your company on the importance of CX and how your corporate culture can play into the success or failure of your transformation. Without the CX at the center of your business, metrics and perceived "successes" are likely for the benefit of your business' bottom line — cost/profit management — versus truly serving the customers' needs and desires.

We have seen this many times over the past 25+ years. Hundreds of customers found their CX lacking. Without change, companies will continue to experience the same outcomes and repeat the same mistakes that have frustrated passengers in all industry verticals for decades.

Companies share their concerns with two main areas: not delivering on promises from internal resources and the constraints of technology.

Based on those two areas of concern we often hear statements like:

- "I want it in the channels of my choice." (internal resource)
- "Instead of sending notifications for me to take action, you should know what I need before I do — and do it for me!" (internal resource)

- "We are missing features our competitors have." (constraints of technology)
- "We know we need it, but the cost is too high." (constraints of technology)

For decades it has been the same story because the measurements for success have remained the same with little to no acknowledgment regarding the evolution of customer needs.

In our debut book, you will delve into the ever-changing landscape of customer expectations, exploring the importance of understanding consumer behavior and leveraging technology to anticipate and meet their needs. You will witness the birth of a customer-centric culture, where organizations embrace empathy, agility and innovation to craft experiences that leave a lasting impact. Lastly, everything will be set for the transformative journey that lies ahead of you.

## Five key themes

Enterprises see themselves as the nucleus. Quite simply, that's wrong: the customer is. When you endeavor to change your culture to support that thought process, you will see an improvement in CX.

When you finish the book, you will walk away with five key themes:

- Ownership and an aligned focus is important to deliver CX
- Leveraging metrics of meaning can create a new baseline
- Focusing on desired outcomes is more important than leading with tools and technology
- The decision to make your customer the nucleus of your company is paramount
- A plan for continuous improvement must be part of your culture

Essentially, your view of your customer experience will forever be changed.

## Who Are We?

We have over 50 years of combined experience (yes, we are that old) in the contact center space, holding roles ranging from agent to executive. Chad, with his deep knowledge of customer-centric strategies, and Gina, with her keen understanding of emotional connection and business growth, form an unstoppable duo committed to unlocking the true potential of customer experiences.

Chad reveals the power of digital transformation and data-driven insights in creating personalized and memorable interactions, while Gina emphasizes the significance of emotional connections in fostering lasting relationships and improved employee experiences.

As operational consultants, we have engaged with over 100 enterprise organizations. We have experienced or seen almost everything that can occur in the CX space. It is what inspired us to write this book. We continuously saw our customers running into the same issues. Only the select few who chose to build their business around their customer as the center of their enterprise are excelling in the industry today.

We have spent many an evening over a cocktail or two sharing stories from our numerous ventures within the contact center space. The more we discussed the scars, successes and lessons learned, the more we realized we were aligned in our quest to help readers like you see the benefits to be had by focusing on the CX and Metrics of Meaning.

## Who This Book Is For

If you are responsible in any way, shape, or form for how your company interacts with your customers and any of the following issues sound familiar to you, then this book is a must-read:

- Budget constraints inhibiting innovation
- Corporate/business/IT goals are not aligned
- Technology is so out of date; our main focus is to "keep the lights on"
- Inability to deliver consistent customer experiences across channels
- Uncertainty of measurements to define success and business value
- Total Cost of Ownership (TCO) / Return on Investment (ROI) takes precedence over CX
- Owner of CX isn't clear
- Monitoring our performance and the technology is a much higher priority than modifying how our customers interact with us
- Multiple tools/technology are being used to fix issues without any success
- The changing industry makes it hard to keep up with the status quo service trends
- Continuous improvement isn't baked into how we do business

As you read through the following pages, you may disagree with some points, or think others aren't applicable to you. That is okay. We ask that you keep an open mind. At times, you may be tempted to dismiss the ideas as you may think, “That’s not how we operate” or “This can’t work.” In those cases, we challenge you to read on and think “What if?”

We hope you will read this book and consider strategies that enable you to achieve continuous improvement of your customer experience.

Now is the time: you can’t win a race you haven’t started. Your customers are evolving; are you keeping up?

We invite readers to join us as we navigate the complexities of the customer experience landscape, offering practical strategies, thought-provoking insights and real-world examples that inspire and empower businesses to embark on their own quests for customer-centric excellence.

The adventure begins, and the pages ahead hold a wealth of knowledge, guiding readers toward a future where businesses thrive and lives are changed through your embracement of the transformative power of customer experiences.

Start where you are and get ready to embark on this enlightening voyage.

★ ★ ★ ★ ★

# CHAPTER 1

## The Foundations of CX

*"If you're competitor-focused, you have to wait until there is a competitor doing something. Being customer-focused allows you to be more pioneering."*
*—Jeff Bezos*

No story illustrates this point more than the one shared by Simon Sinek in his book *The Infinite Game*[1] about two of our generation's biggest competitors, Apple and Microsoft, and why it's so important to focus on your customers instead of on your competition.

Simon attended an education summit for Microsoft. He was gifted the new Zune, Microsoft's version of an MP3 player, set to compete with Apple's iPod.

The Zune was elegantly designed. "Quite exceptional," Sinek writes. As he shared a cab ride back to his hotel with a senior Apple executive, he thought he'd share his thoughts on Microsoft's new Zune MP3 player. He turned to him and said, "You know, I spoke at Mic-

---

1 Simon Sinek, *The Infinite Game* (New York: Portfolio Penguin, 2019)

rosoft and they gave me their new Zune, and I have to tell you, it is *so much better* than your iPod Touch."

The Apple executive's response? "I have no doubt."

This was atypical. How was this Apple executive so unfazed that their competitor had a much better product than his own company had — and he knew it?

The Apple exec. didn't care that the competitor had a better product because he was aware of the bigger picture — not concerned with achieving an immediate "win" over their competitor. He knew Microsoft would have some better products, but so would they. We all know the end of this story: the Zune and iPod became obsolete with the introduction of the iPhone.

Let's walk through a scenario we experienced in the field.

During a week-long series of workshop sessions with an insurance client's IT, Data Sciences, and contact center teams, we ran into one of the most interesting outcomes of our careers. Our mutual objective was to understand current customer experience practices, challenges and opportunities. We'll give you a bird's-eye view of the situation:

> Boom! There it was. The hit in the head that every consultant hopes never happens, but always anticipates coming at the worst possible time. And, it was *now*.
>
> "Good morning! Let's do a stand-up. I mean, actually stand up, please!" said Chad.
>
> The room of client participants begrudgingly pulled away from their laptops and stood up, looking somewhat curious but mostly inconvenienced. Not a good sign.
>
> "Okay, today is the final day of our workshop series," said Gina. "As we planned with the agenda, today is all about 'Art of the Possible,' to apply what we learned this week and challenge

ourselves to come up with ideas to explore. Forget about time, money, resources. Think *big*!"

The room was still silent, and to add to the mood of the crowd, frowns were appearing on a couple of faces.

Chad decided to engage one such frowning team member. "John, I can't help but notice that you look less-than-positive about this final phase of the workshop."

Now that the elephant in the room was being called out, the discussion could get "real" at a new level.

Chad continued, "We all want the same thing: to find new ways to innovate, conduct Proof Of Concepts (POCs), fail fast, refactor, achieve minimum viable product (MVP), and then enter continuous improvement. One of the best approaches is to work backward. However, this workshop is for you and the team, so I want to be sure we are in sync before we start the session today. Will you please share your thoughts? Remember: this is a safe forum where all ideas and thoughts can be shared!"

John looked slightly relieved, paused a moment, and said, "Well, all week we have been talking about challenging ourselves to think big, don't paint ourselves into corners by not anticipating larger scope, and what MVP and continuous improvement are. I get it. Heck, we all do! But I don't know if we can actually *do* it."

We shared a glance across the room. Finally, we were getting some honesty out of the team. John's concerns were certainly shared by others in the session. But we couldn't address them until one of the active participants voiced it out loud.

Appealing for support and validation, John looked around the room at his co-attendees, then resumed. "I mean, we are talking about customer experience, not creating software from scratch. We have learned so much this week, and I think I speak for us all when I say we really enjoyed the sessions with you and Gina. But our goal coming into the workshop was always about reducing costs. If we can't do that, I don't see how we can possibly do as you said and 'think big.'"

Heads nodded.

Gina said, "John, I appreciate your trust to provide such a candid response. We would expect this, as it is quite a departure from what you do on a daily basis. How about we address this in terms of crawl, walk, run? We are looking for you to state your nirvana. Mechanically, we understand it takes a much more measured approach."

After looking each attendee in the eye one by one, she continued, "John and team, our goal is to get you to establish new metrics of meaning and hyperfocus on your customer experience to achieve business outcomes such as reducing costs. We will work with you to address both your cost and delivery concerns one step at a time."

Almost instantly the team felt reassured that we heard them, and there was an understanding of the gravity of the work ahead of them, but the path now seemed possible ...

Over the last five years, we have consulted with multiple contact centers of every size in a variety of verticals. Invariably, the number one focus has been cost reduction. Yes, the companies all wanted to ensure great customer experiences and adopt innovative solutions such as transcription, translation, sentiment analysis and more. But when it came down to execution, operational teams gravitated right back to managing their contact centers for efficiency with only the basic customer satisfaction metrics to gauge their effectiveness.

As we started the engagement, we took what we felt was a logical, easy starting path:

"Please share with us what metrics you use to measure your customers' experiences," Chad said.

The response was as expected. "Oh, we use CSAT from post-call surveys."

*That's a great start,* we thought. So, we started to dive deeper into the topic.

When we asked how the contact center management knows a customer interaction is going poorly in real-time or a supervisor is needed right away, we were not surprised to hear "Agents stand up and wave at their supervisor, and the supervisor can then listen to the call or coach the agent."

That's a reasonable process, given that most contact centers do not have AI/ML solutions for real-time evaluation of calls and automating notifications to supervisors.

However, when we asked what their plan was to evolve to AI-based solutions and to address the fact that more and more agents work outside the contact center and can no longer simply wave their hands for help, the response was, "We do not have the time or money to commit to such projects. We'd love to, but ..."

While we want CX to be the nucleus of your business, in some scenarios, that can't happen. In this case, cost can be a barrier of entry to enhance solutions, leaving teams to believe it's not realistic. For the duration of our engagements, we documented teams' focus on cost-specific metrics versus really measuring and acting on the outcomes delivered to customers. In the course of the read-out, the management teams nodded their heads, said they agreed with the assessment, and admitted that our recommendations for improvement were "spot on," but they would not be able to address them "in the next couple of years."

There it was, again. The hit in the head. Despite winning the trust of the customers and gaining their validation that our recommendations were great for their challenges, the reality was it was not going to happen now or in the near future because they didn't allocate budget to customer experience.

## Seven Key Factors That Impact Overall CX

When it comes to CX, you need to consider the seven factors that impact it.

### TCO/ROI taking precedence over CX

People are focused on technology, costs, business-centric metrics and TCO / ROI —pretty much everything except CX. We have worked at companies where the CX was impacted by delays accessing backend systems, and the fix was never implemented because of cost and internal disputes between IT and business owners about who would pay for it. In another case, we experienced a focus on bringing in all the latest and newest technology for cost savings without any thought to how this would impact their CX. In almost every case, cost savings take precedence over CX enhancements.

Many of our customers have implemented a new solution focusing on reducing AHT to reduce agent headcount rather than to ensure the customer gets *all* their needs taken care of in the same interaction. They believe lower AHT = lower costs, without considering the customers' frustration in having to make multiple calls if they can't get a resolution on the first call. Companies find it challenging to invest purely to deliver superior customer experiences without offsetting costs by reducing time and money elsewhere. Thus, the TCO and ROI take precedence over the CX focus.

We understand that businesses are being asked to do more with less; business constraints are a reality and, ultimately, we usually witness the customer suffering.

In Chapter 3, we will talk about aligning metrics to the CX.

## Corporate culture

Some customers start to go through the transformation of focusing on CX and end up stopping for various reasons. Each time this occurs, we witness their culture playing a significant role in their project abandonment, leaving many of their improvements to become shelfware. We have all seen absolutely brilliant initiatives killed thanks to passive-aggressive approaches of those required to operationalize. In one case in particular, the C-suite stated they supported the initiative and wanted to make the customer a priority. Yet when it came down to actually changing behaviors, the executives were unwilling to change their own mindset. Within three months, their direct reports saw their actions not matching their words, and soon they lost the drive to continue on with making the changes required to become better. Status quo became acceptable, once again, at all levels within the organization. It was too hard to change from the bottom up, so the initiative died! We challenge you not to let your organization become a statistic but rather stand out among the crowd.

It is a known fact corporate culture plays an essential role in both your CX and your employee experience (EX). We have worked with many clients in the field that have their vision statement and values on the wall. The intent is good, with the customer being put first. However, the company's actions often struggle to live up to them. Commonly, in situations like this, we see high employee turnover. We have witnessed an invariable and direct correlation between EX and CX ratings.

## *What's in your petri dish?*

Your culture can only grow what is in your "petri dish." Imagine a clean petri dish in a lab. It is sterile, barren, and void of any life. Technicians very carefully introduce only what they want to be grown and observed. No other "contaminants" may be introduced that would invalidate the controlled culture. Running an enterprise is not as steady and clean! Your culture yields all the intended, good elements, undesirable traits and everything in between. Without taking stock of what is in your petri dish, it will be hard to appreciate what is essential to keep and what may need to change over time.

Cultures change as people come and go, making it more challenging to establish the perfect environment. However, culture starts from the top, and every effort to bring new members into the cultural beliefs, practices and leadership principles should be a primary focus.

*Figure 1 Dall-E2, OpenAI, openai.com/dall-e-2*

## Communication

A recurring challenge for companies of any size with regards to focusing on the end customers over their driving, daily business needs is communication. The best-laid plans tend to fall apart once teams start their work for the day. Actions to ensure the CX is the first priority are rarely implemented because of the competing priorities to simply run the business as efficiently as possible.

While working with a client, one of the lead developers for web applications told us her major frustration was, "Business owners come to me and ask when the website updates for new products will be done. More times than not, this is the first I hear about the changes! IT does not bring us in until the business escalates, and then we are expected to deliver immediate changes. It's embarrassing when customers complain about errors we did not catch!"

Another example of miscommunication is commonly introduced by marketing launching a campaign without notifying all parties involved in the customer life cycle. While consulting for a large telecommunications services provider, we reviewed their campaign outcomes reports for the last 30 days. The reports indicated large outbound campaign activity spikes, resulting in an increase of inbound calls to the contact center.

We asked what the concluded cause and effect were: "Oh, those were due to marketing campaigns for the latest release of our phone going out, and our subscribers wanted to talk to agents about details, plan changes and how long it would take to get the phones."

Then, we probed a little deeper and asked, "Did your outbound messaging anticipate those questions and include links to answer them?"

The conversation yielded that the marketing department launched the campaigns "in a vacuum" and did not collaborate with business owners, IT or contact center teams. There was no opportunity for input on the messaging and readiness for customer inquiries after receiving the marketing messages.

Had they consulted with the contact center leaders, they would have heard something to the effect of, "I am glad you asked for our input. For the last three months, your team has been blindsiding us with outbound marketing, leading to inbound calls that we are not staffed or trained to handle. This frustrates our customers and agents and leaves our customers doubting we know what we are doing."

These are not isolated examples showing how a lack of communication directly impacts the end customer. Each of your business units has its own metrics and will achieve them, regardless of impact on the customer. This is a phenomenon known as WYMIWYG, which will be discussed later in Chapter 3. You could probably name an example of competing business drivers impacting your company's customer experience off the top of your head. The breakdown in communication is because of this singular focus. We're sure it, too, was avoidable. The big question is: Have any measures been put in place to mitigate such communication challenges from affecting customer experience in the future?

## Lack of clear CX ownership

One of the first questions we ask our clients is, "Who *owns* your customer experience?" More often than not, the answer is, "We really don't know" or "Everyone." In either case, it means no one owns it. Only in the last five years have we seen the advent of Customer Experience Officer (CXO) and comparable titles in roughly half of the companies we have worked with. Where a CXO is in place, two scenarios emerge: the CXO is a one-man-band with no staff to deliver on the CX vision and strategy, or there is a full CX team in place but operating in parallel to the contact center/service delivery teams versus as a cohesive group. In all cases, we have observed challenges with making meaningful CX changes due to a lack of resources, the ability to assign accountability to other teams to implement and competing priorities. Without a clear owner, we have always seen chaotic, unpredictable outcomes.

Let's take a services company. It's their job to provide top-notch CX deliverables to their clients. However as the projects start, confusion ensues about who owns the actual experience of the customer. Is it the sales member who sold it? The professional services team who delivers it? The customer success manager who claims the relationship post-implementation? Or technical support who comes into play post-implementation as well? Each of these teams has a singular focus of deliverables, and everyone loses track of what the ultimate goal is: trying to improve their customers' experience. Where there is no clear ownership of the overall CX or top-down support, there is usually a lack of focused vision and strategy to achieve the desired CX outcomes.

## Lack of innovation

*"Necessity is the mother of invention."*
*—Plato.*

While we absolutely believe and adhere to Plato's words when needs arise, we need to continually think big and innovate beyond necessity. Let's walk through this by addressing a statement we commonly hear: "We have always done things this way."

Resistance to change is compounded by legacy mindsets in general, causing them to stay stuck. This happens all the time: relying on the same reports decade after decade, taking little to no meaningful action based on the report data, and generally depending on what we refer to as Metrics of Convention versus Metrics of Meaning. The high-level observation is that organizations do the same thing while expecting different outcomes. We understand that part of the reason may be attributed to budgetary constraints, and part is fear of change. This results in teams lacking the resources to disrupt current processes and drive innovation. Repeating the same things and expecting new outcomes is not likely to net new, enhanced CX outcomes.

Most enterprises we have worked with fundamentally lack innovation relative to CX enhancements. From the largest financial institutions to the most technical software companies, development approaches for contact center solutions have not been geared to iterate quickly, fail fast and evolve to production-ready status in lockstep with customer expectations and needs. Other business units of such companies have adopted Agile and DevOps and are methodically delivering business value. Contact center teams have started to adopt the same practices, but they are playing catch-up and still cannot innovate at a rapid pace. This leaves CX behind with lengthy gates of new functionality being introduced in 6, 12, or 18+ month timelines.

We also see an increase in third-party engagement, and in most cases, the internal team has too much history and baggage to assess and implement what is required to grow to the next level. We are seeing enterprise-level companies think outside the box and incorporate different learning techniques to upskill their teams in today's competitive market. The importance of the CX isn't a question for most people anymore; how to fix it is where the majority of discussion is occurring today. We have found our customers at various stages in the process, sometimes all at once. Some are in the foundational area where they just need to get their day-to-day operations working, improving the CX for their customers. Others are working on refining, enhancing and optimizing what they have to provide the next level of service. The crucial thing to remember is: start where you are. All the puzzle pieces have to fit in the right place; unfortunately, you have a problem if they don't.

The CX is key to having a successful business. Put your customers at the forefront and ensure they are happy, and you're golden. That's how many of the biggest corporations today evolved from garage start-ups to globally iconic brands: they valued the customer over everything else.

## Technology

Another common observation in the field is that companies feel they can't focus on the customer because they are too focused on keeping outdated technology up and working to provide basic service to their customers. While a lack of technology has been known to contribute to a poor CX in the past, we want to be very clear that all the technology in the world isn't going to fix your CX problems. Don't get us wrong! Parts of technology can definitely impact the experience, but it is highly unlikely it is the root cause of the issue. All the best technology may net nothing without the right application.

With technology playing a part in the CX, ensuring the business and IT goals are aligned with enterprise is essential. We haven't had a client yet that didn't experience the divide between business and IT, and this impacts your end-customer experience and often your EX.

Thus, it is essential to set a common goal and focus for the company to align at the enterprise level, with the focus being on your CX. Technology decisions should be based on how they will enhance the end customer's experience.

## Workflow

We come across many flavors of workflow management, software life cycle development and IT frameworks. Each has its own set of strengths and weaknesses. We have seen Agile/DevOps, Waterfall and "Wagile" practices deliver remarkable results. On the flip side, we have seen teams struggle regardless of the process they follow. These observations have enabled us to be objective: put the CX first, use the development and delivery method your organization embraces, and iterate as quickly, accurately and effectively as possible with the end goal in mind at all times. We understand that change takes time, is often uncomfortable and is rarely easy. However, if you are focused on CX, are open-minded, can embrace whatever methodology will work for your team and can leave the politics aside, you can implement any change successfully. Workflow is rarely the inhibitor to your success.

## Call to Action

We know you are saying, "Okay, sure we have issues; everyone does." You are right! But now you need to identify your issues and know your baseline so you can better measure success. As you embark on your journey, we ask you to take some time to assess where you register within the following areas.

TCO/ROI Takes Precedence over CX:

- Are you driven by ROI vs. CX?

Corporate Culture:

- What are the strengths and weaknesses of your corporate culture?
- Is there support for change starting at the top of your organization?

Communication:

- Where is the biggest communication breakdown in your company?
- What are your suspected root causes of the communication breakdown?

No clear CX ownership:

- Do you have a clear owner of CX or a plan to improve the CX in your organization?

Lack of innovation:

- Are you measuring the same metrics you were 3, 5, or 10 years ago?

Technology:

- Is technology interfering with delivering outstanding CX?

Workflow:

- Do you have an established workflow or process that everyone aligns to?

Now that you've assessed where you're at, you're left wondering how your company can improve. One powerful way is through engaging in business simulations, the topic of the next chapter.

★ ★ ★ ★ ★

# CHAPTER 2

# Using Business Simulations to Effect Change

*"I hear, and I forget. I see, and I remember. I do, and I understand."*
*—unknown source*

In 1963, Samuel Eilon expounded on the following three major purposes of business simulations:

- To provide a rounded view of the tactical operations at the corporate level
- To be used as another tool for training and allow the players to confront the various impacts of their decisions
- To feign market trends to improve the player's capacity to face changes successfully[2]

---

2 Quoted in "Business Simulation Games," StudiousGuy, accessed August 2023, https://studiousguy.com/business-simulation-games/.

Before performing their real-world duties, military commanders, doctors, pilots and emergency response teams undergo simulations. As a result, individuals are better prepared to deal with not only the day-to-day tasks of their roles but also unforeseen issues. For those of you who haven't had the opportunity to participate in a business simulation, it is one of the best ways to facilitate people working together and aids in the gradual reformation of a company's culture to adopt change. When you get a group of executives, IT, business owners and support personnel all aligned to common outcomes of a business simulation, it is astounding how much empathy is established in such a short time.

If you take nothing else from this book, you need to know you have options to help you begin the culture change that may be required to begin putting your CX first.

## Why Business Simulations

To improve CX, businesses invest a lot of money in employee training. But afterward, they frequently fail to see the predicted behavioral and quality changes. The fact is that corporations no longer rely on multi-day training programs. Yet today's businesses must continually invest in their staff. With the Great Resignation happening, demonstrating your commitment to your team's work environment being productive and healthy could make a difference in retaining your top talent.

Instead of asking your employees to commit to multiple days in training sessions together, focus on a single-day business simulation. In one day you can empower your team to have a higher degree of empathy and an increased appreciation for the need for transparency and collaboration to achieve common goals.

These simulations are making an impact in the industry for several different reasons, primarily to facilitate communication, improve collaboration and help build an empathetic culture where people want to show up to work.

In one of our engagements, our client was the vice president of customer care in the healthcare industry. Imagine walking into a meticulously neat, organized, well-decorated office. Bona fides of considerable prestige were precisely posted on the walls. An air of "down to business" and "you had better not be wasting my time" wafted into our noses as the epiphany occurred: "This executive is the *real deal*."

Swallowing hard, looking at each other and committing to move forward with our proposed agenda, we walked in, introduced ourselves and went through our proposal for conducting a series of business simulations. The objective was to enable teams within the customer care department to establish empathy and learn how to work together to solve common problems. At the end, there was a really uncomfortable pause before hearing the verdict.

"When I first heard your pitch about business simulations, I thought it was gimmicky. Almost instantly, I closed my mind to the concept and thought there was no way I was going to pull my team away from their day jobs to spend a day in a simulation. After seeing the video clips from your previous sessions with clients, I changed my mind."

There was a palpable ease of tension on our part. After all, she had not said no ... yet!

The VP of customer care provided us with several conditions on what it would take to do a pilot of three business simulation sessions. We diligently worked on the proposal, prepared a timeline and presented it, and it was approved.

After the pilot completed, we once more entered the picture-perfect office, and, surprisingly, were warmly welcomed with a wide smile far different from the initial, clinical feeling of our first meeting.

The customer care executive said, “Now that we have conducted three simulations, I am convinced that we need this to be an ongoing program in order to help us change our culture and rebuild as a team, from top to bottom. You have convinced me that what I think is operating perfectly is *not,* and all the feedback from my team has been around how amazing it was to understand the impact each has on the other, regardless of their roles.”

Even more surprising, she called out anecdotes and testimonials from her team. Our mouths nearly dropped open. For her to have taken this much personal time to organize, review and appreciate was exceptional.

Clearing her throat, she proceeded to share specific feedback.

One of the developers said, “Oh! What I can see us doing better on a daily basis is having empathy for each other. We all have far more work than we can complete, and it is very easy to get frustrated when others can’t stop what they are doing and complete what I need right now. Going through this simulation made us all realize we are in this together, none of us can complete everything alone, and we need to find ways to collaborate and achieve goals as a team versus ‘They did not complete on time, so I can’t be held accountable.’”

An IT executive responded with, “She is right. I almost always overstate timeframes for delivery because we can’t complete what we have, much less what is coming in on a daily basis. Going through this session together has made me appreciate the impact our decisions have on other teams. If we all are able to let down our guard and work together to build processes where everyone's needs and abilities to deliver are known and appreciated, we can prioritize, commit and deliver with much better success."

After going through a handful of such items, she looked at us, smiled, reached out her hand, shook each of our hands and thanked us for changing not only the teams' perspectives but hers.

"Your statement about culture starting at the top was true. I denied it at first, and it took me a long time of reviewing the pilot data and interviewing participants myself before I could accept it. Thank you for making me aware."

## Benefits of Using Simulations

Many people learn by doing. In the past few years, with the increased adoption of Agile and DevOps, we have found teams yearning for something that helps address the communication and collaboration breakdowns that companies often suffer.

A business simulation enables participants through an interactive learning experience that aims to imitate real-life workplace scenarios. This is accomplished through:

- Practicing and developing communication
- Financial analysis
- Strategic thinking
- Teamwork
- Learning to provide constructive feedback
- Market analysis skills

Business simulations enable team participation and assist your business in developing effective alignment toward your strategic goals.

The benefits of a business simulation are vast. A few of the key takeaways include:

**Experiential learning for accelerated performance:** Experiential learning is one of the most successful methods of team training.

**Developing important skills in a realistic setting:** Participants improve their abilities by participating in real-world simulations. To mention a few, these abilities include teamwork, critical thinking, communication, and comprehending value.

**Collaboration can help you learn more:** Interaction, collaboration and teamwork with peers are some of the most effective methods for achieving results. Recognizing how the workflow affects the various divisions within the organization is essential to learning from the simulations.

**Encourage unprecedented participation:** Simulations, according to research, boost student participation in the learning process. When relationships, cooperation and gaming aspects are combined in one package, it helps create an atmosphere that participants desire to experience.

**Enabling organizations to align their employees on strategy and communication:** This has been the most beneficial outcome we have witnessed. Successful business simulations are tailored to address your issues and bring out the behaviors you want to change within your organization in a fun, hands-on way. In this way, you can give the participants real-world experience to take back and apply to their day-to-day roles.

Business experience is about improving delivery via experiential learning and engaging activities.

We currently conduct three different simulations for our customers, ranging from Agile/DevOps, ITIL, and SAFe principles. No matter where your organization is in your cultural or communication assessment, these specific simulations will help. Typical learning outcomes include:

- Identify how to implement customer-centric initiatives across all departments
- Discover how to design with the final goal in mind (Teams, for instance, study how to map value streams and target the removal of obstacles)
- Learn to accept responsibility for outcomes
- Recognize the significance of providing feedback to one another and working in a culture where failing fast is encouraged
- Establish a culture of ongoing learning and development
- Realize the need to automate whatever they can
- Promote efficient communication and cooperation across distributed teams

## Simulation Case Study

Let's go back to the insurance company from Chapter 1 where we evaluated the clients' current CX practices. We left the series of workshops with a CX transformation plan for the customer to execute. When we returned for a touchpoint three months later, we found that the culture was operating exactly how it had in the past. They had addressed the technical issues they were facing and had an increased focus on ROI. However, they still had CSAT issues, high staff turnover and low employee satisfaction ratings. As we sat over dinner with the C-suite executives, listening to the issues they were facing, we heard loud and clear that they were ready to focus on the culture. They had no idea where to start or how to get the teams on the same page, but they certainly identified their major hurdles:

- Communication
- Cross-functional Collaboration
- Customer Experience
- Workflow

We discussed how a business simulation could jump-start the team, and the executives half-heartedly agreed. There were definitely some skeptics at the table.

Two weeks later, we kicked off the first business simulation with participants including C-level, developers, managers and business representatives. The simulations were structured with each participant taking on a character/role.

The first simulation participants naturally, without speaking, broke into the business and IT sides of the room despite us saying, "It does not matter what your role is in the real world. For this simulation, anyone can play any role." Inevitably, the room divided, and people gravitated toward the roles most closely aligned with what they did daily.

After the first simulation, we worked with our sponsors to predetermine the roles of the subsequent participants. Every participant gains a genuine appreciation when they "play" a role they otherwise don't get much exposure to during their typical workday. After the eight-hour session where different roles are assigned, higher adoption rates and a higher percentage of sustained behavioral changes occur. It is also crucial for leadership to take a "doer" role to understand how their clear communication on the overall corporate goals truly impacts and empowers their teams.

Note: Because the first simulation produced results even the executives couldn't believe, we completed nine more within the company, encompassing over 100 team members. We maintained the model from the first simulation, having participants ranging across the entire employee hierarchy.

In these sessions, as we set the stage for how important communication is to the session's success, everyone nodded and confirmed they appreciated and understood the point. Then the simulation began. Communications were chaotic, there was mass confusion on how to execute the simulation despite the clear instructions, and the first round typically ended with frustration and categorical looks conveying, "Wow. We suck!" or "This game is stupid and a waste of time." At the reflection part of the round, the team pointed the finger in all directions. Yet every time, the true culprit boiled down to business needs versus IT's technical debt and capacity constraints. Sound familiar?

After the first reflection, the overall agreement from the team was that they needed to improve communication. The second round began. This round focused on deciding what work would be done and when and how they would make the decisions. Again in the first simulation and subsequent sessions, the company's culture was apparent — everyone worked in their silo, so very little had been accomplished at the end of the round.

Frustration grew, and in the reflection part of the round, when it came to light that one role had the answer they had been looking for but that individual hadn't spoken up, we heard, "Well, that is so typical of what happens every day around here." Two of the greatest benefits are that the simulation addresses real-life issues and the participants get to see the impact of their actions or inactions in real-time.

As the third round finished, the teams learned they needed to establish procedural communications. More specifically they appreciated the need for:

- A repeatable process to review changes
- Establishing a prioritized backlog
- Commitment
- Executing their committed tasks with an awareness of the impact on their teammates

They also learned to have *empathy* for one another. In one of those simulations, a CTO stated, "Brian, I am sorry I pointed the finger at you in that first round. Now that I see what you are up against with five of us loading you up with work, you can't succeed! This simulation makes me aware of how unrealistic my expectations are." The lightbulb went on above his head. The best part is that this didn't just happen once and only to the CTO. There isn't one simulation we have conducted where someone doesn't have the "AHA" moment.

As that first simulation concluded, we saw the team uniting, people helping one another, regardless of their assigned roles, and everything executing like a well-oiled machine. The mood in the room became positive; everyone was over-communicating, and the energy in the air was contagious. As facilitators, we had to gradually rein the group back in and set the stage to debrief the outcomes.

- What did not work at first? Why?
- What did you refactor and do differently?
- How did that lead to improvement?
- Do you think there are still areas for improvement?

This line of open-ended questioning drove lengthy dialogue. After 30 to 40 minutes, the conclusions from all the simulations were very similar and included the following observations:

- The importance of being aligned and communicating openly
- The need to demonstrate empathy to internal resources and customers
- A greater appreciation for the importance of visible work in collaboration efforts
- Commitment to making small changes in how you operate today greatly impacts both EX and CX
- Stronger understanding and commitment to the overall corporate strategic approach

This is comparable, on a smaller scale, to full culture change. Many companies have made business simulations a part of their corporate training program. Some of our customers have monthly/quarterly coaching sessions, some run annual refresher workshops and others run a simulation each time a new scrum team develops.

## Call to Action

- How do you see this process adding value to your company?
- What specific issues would you want to draw out for your teams to resolve in the simulation?
- Are you ready to see how a day-long investment can improve your CX and EX?

As you can see, a simulation can help you down an alternative path, but in doing so, you need to avoid the common pitfalls outlined in the next chapter.

★ ★ ★ ★ ★

# CHAPTER 3

## Best Practices and Common Pitfalls

*"What gets measured gets managed — even when it's pointless to measure and manage it and even if it harms the purpose of the organization to do so."*
*—Unknown Source*

Let's focus on three of the top pitfalls that, in our experience, you should work to avoid. The content in the rest of the chapter applies to all three and covers a broader context to allow you to consider other pitfalls unique to your CX environment.

- Pitfall 1: Measuring the wrong thing
- Pitfall 2: Measuring Frequently, Acting Infrequently
- Pitfall 3: Doing what you've always done and expecting different results

*Figure 2 David Crane, Pitfall!, Activision, Atari Video Computer System, 1982.*

Many of you may recognize the above picture from the 1982 Atari game *Pitfall*. We feel this is a perfect visual representation of what we are going to cover in this chapter. Imagine the water to be the reports and practices your business has been embracing for the past few decades.

## Pitfall #1: Measuring the Wrong Thing

In 1956, V. F. Ridgway published a paper, "Dysfunctional Consequences of Performance Measurements," criticizing the measurement mantra.[3] Simon Caulkin, a columnist, neatly summarized Ridgway's argument that he made in his paper: "Ridgway clearly was onto some-

3 VF Ridgway, "Dysfunctional Consequences of Performance Measurements," Administrative Science Quarterly 1, no. 2 (1956): 240–47. https://doi.org/10.2307/2390989.

thing. Not everything that matters can be measured. Not everything that we can measure matters."[4]

The latter part of this statement caused us to pause: if you keep measuring metrics that do not deliver the CX outcomes needed, you will fall right into the first pitfall of measuring the wrong things.

Not long ago, Chad attended two in-person training courses: Scaled Agile Framework (SAFe) and DevOps Practitioner. His instructor, Ray, introduced the class to a truly magical acronym that shook his foundational understanding and appreciation of the conventional metrics used by contact centers: WYMIWYG. (No, that is not a typo of WYSIWYG, in case your mind jumped where his did when he saw this drawn on the whiteboard.) It actually means What You Measure Is What You Get (WYMIWYG).

Ray explained that the acronym is akin to a self-fulfilling prophecy for enterprises, for whatever metrics your business decides are the most important to measure will be the sole focus of the entire company, down to individuals. You will almost certainly hit those metrics at any cost, even if teams or people know they are not fully aligned with the actual CX delivery desired.

Everyone is motivated to work toward them, especially key individuals who often have their performance and compensation associated with these metrics. Suppose AHT is what drives all the decisions on the floor in your contact center. In that case, all other metrics are a distant second or lower priority to attain. One way or another, your

---

4 Quoted in Anne-Laure Le Cunff, "The Fallacy of 'What Gets Measured Gets Managed,'" NessLabs, accessed August 2023, https://nesslabs.com/what-gets-measured-gets-managed.

entire focus is to meet or beat the AHT threshold. But if you're measuring the wrong thing, your CX gets left behind.

To set the stage for the next section, we ask you to put on your contact center Director/Manager/Supervisor hat!

If you've spent time in the contact center industry, you'll notice that the metrics most companies use to assess contact center performance haven't changed much over the years. While vast improvements have been made with more channels, customer journey ideas, AI, and sentiment analysis, we find the metrics required to highlight the value of these changes are not apparent. There has been no widespread shift away from using the same data to achieve the same outcomes when measuring contact center performance.

Naturally, CEOs seeking to improve their firms' competitiveness through better CX have been investing significantly in a variety of metrics:

- Mapping and tracking the customer's journey
- Determining whether or not to advertise
- Judging how well a firm serves its customers
- Predictive modeling of CX journeys
- Determining business outcomes

This is a departure from Metrics of Convention and starts to move organizations in the right direction: measuring what really matters. However, as CEOs set these goals to establish new , it's difficult for the operational teams to quantify, thus resulting in falling back into old habits of measuring what they always have.

In our work, we've noticed that more and more businesses are dissatisfied with their current reporting systems and want to develop a measurement strategy that integrates all of their customer experience indicators into a single model to achieve a customer-centric approach. In terms of design and execution, a comprehensive approach to evaluating CX may take various shapes.

## Pitfall #2: Measuring Frequently, Acting Infrequently

We have had the opportunity to work with customers who measure at the right interval, however they did not act on customer feedback in a timely manner.

> Sitting with the operations manager one day discussing measuring CSAT, we could hear the pride in his voice. "We set up a cadence, we are sticking to it, and we solicit feedback from our customers at four points through the customer journey." He listed them out for us:
>
> - After the sale is complete
> - After Discovery has occurred
> - After Implementation is completed
> - Once a quarter after project implementation
>
> We were eager to dive into the magnitude of data they had collected. As we started sifting through responses, we noticed a common theme: post-implementation, scores were dropping at an alarming rate. The sentiment became very negative: "Please stop sending us these surveys if you aren't doing anything to fix the problem."

When we asked the operations manager if he knew what the customer was referring to and if he could explain it to us, again he replied with confidence.

"Well, our customers ask us to fix things, but we are so far behind in our release schedule that it's been eight months since we've launched any new functionality, or fixed the existing issues we have. Our customers are really upset!"

"That makes sense," we replied cautiously. "Can you help us understand why you continue to solicit feedback from them when you haven't acknowledged their previous responses?"

"I'm given bonuses based on how many surveys I send out and the number of responses," he meekly replied. We could see the alarm on the faces of the leaders in the room.

"We totally understand. However, what you do with the information is just as important as measuring your metrics, such as CSAT, at the correct interval. Customers love to be engaged and to feel that they are being heard. However, actioning that feedback has to become a priority for the organization," Gina stated.

We watched the heads of leadership nod and watched the CTO become indignant that "nothing" had been released. We tabled the larger dialogue about why there hadn't been a release for a separate meeting as tempers were starting to flare and the "team" attitude started to fray.

This customer was a perfect case study for us, as they hit all three pitfalls (stay tuned for #3)! They were measuring how many surveys were sent out, so the team was sending out surveys disregarding the fact they hadn't addressed any of the customers' previous feedback. This of course dovetailed into measuring information at the wrong intervals. They continued to survey customers for the sake of hitting metrics, causing them to slide into pitfall #3 like they were stealing home!

## Pitfall #3: Doing What You've Always Done and Expecting Different Results

Organizations often repeat the same things they have done for long periods of time and expect to achieve different outcomes. It is imperative to think deliberately and strategically about shaking up your current practices. A starting point may be OKRs (more on this later). These are two separate but linked measures used to identify a customer's experience.

Examining what you presently measure — how, why and what actions you take as a consequence — is one way to start thinking about success differently. Several clients have recently asked us to formulate strategies to redirect calls to chat or self-service channels during peak call demand periods to provide a better CX. Why wait on hold for an extended period when we can leverage technology to check agent availability for alternate channels, make the appropriate recommendation to the caller and orchestrate the transition seamlessly?

Whenever we bring up the subject of reporting, we're informed that "we need these reports." Our customers either open dashboards and walk us through a voluminous array of "what we use," or they have a list prepared to share. We sometimes get blank looks or perfunctory replies when we question our clients and ask what decisions they make to operate their business based on each report. We dive deep into this area because we know that many organizations have long-standing contact center practices based on the attitude "That's how we've always done it." This mentality is one of the main factors contributing to organizations lacking in innovation.

Often, we find that performance reviews, incentives, and other compensation are directly attributed to the metrics contained in the standard reports, and deviating from them can be costly to contact center personnel and management. When we can objectively evaluate what contact centers are reporting on and how that correlates to running their operations, we find the number of reports required can greatly diminish the overall report count. The challenge we are left with: the metrics used are more focused on employees than the actual CX.

When we met one of our customers, we discovered they had 79 "mandatory reports." By the time we reviewed each report, approximately 20 of them had not been run in over six months. Of the remaining 59 reports deemed necessary, we could distill that list down to 11 reports in total, as many of them didn't offer the information they needed to run their business efficiently. Or, as our review highlighted, the reports only served one purpose: they allowed management to focus on corporate expense minimization rather than how to provide the greatest possible CX.

Sadly little has changed concerning contact center reports. Even though technology has evolved, the application with respect to CX has not.

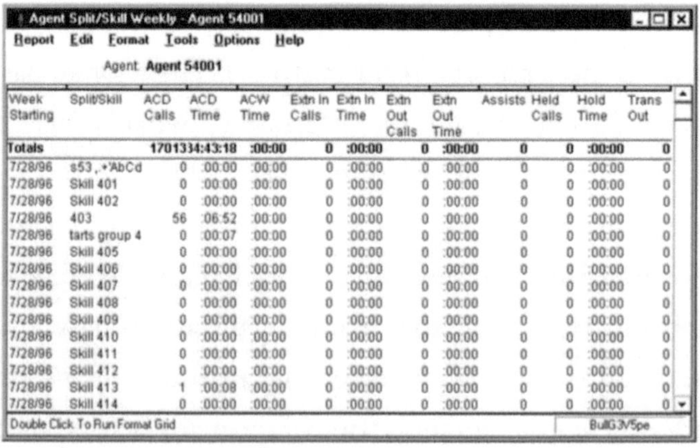

Agent Split/Skill Weekly - Agent 54001

Report Edit Format Tools Options Help

Agent **Agent 54001**

| Week Starting | Split/Skill | ACD Calls | ACD Time | ACW Time | Extn In Calls | Extn In Time | Extn Out Calls | Extn Out Time | Assists | Held Calls | Hold Time | Trans Out |
|---|---|---|---|---|---|---|---|---|---|---|---|---|
| **Totals** | | | **1701334:43:18** | **:00:00** | **0** | **:00:00** | **0** | **:00:00** | **0** | **0** | **:00:00** | **0** |
| 7/28/96 | s53,.+'AbCd | 0 | :00:00 | :00:00 | 0 | :00:00 | 0 | :00:00 | 0 | 0 | :00:00 | 0 |
| 7/28/96 | Skill 401 | 0 | :00:00 | :00:00 | 0 | :00:00 | 0 | :00:00 | 0 | 0 | :00:00 | 0 |
| 7/28/96 | Skill 402 | 0 | :00:00 | :00:00 | 0 | :00:00 | 0 | :00:00 | 0 | 0 | :00:00 | 0 |
| 7/28/96 | 403 | 56 | :06:52 | :00:00 | 0 | :00:00 | 0 | :00:00 | 0 | 0 | :00:00 | 0 |
| 7/28/96 | tarts group 4 | 0 | :00:07 | :00:00 | 0 | :00:00 | 0 | :00:00 | 0 | 0 | :00:00 | 0 |
| 7/28/96 | Skill 405 | 0 | :00:00 | :00:00 | 0 | :00:00 | 0 | :00:00 | 0 | 0 | :00:00 | 0 |
| 7/28/96 | Skill 406 | 0 | :00:00 | :00:00 | 0 | :00:00 | 0 | :00:00 | 0 | 0 | :00:00 | 0 |
| 7/28/96 | Skill 407 | 0 | :00:00 | :00:00 | 0 | :00:00 | 0 | :00:00 | 0 | 0 | :00:00 | 0 |
| 7/28/96 | Skill 408 | 0 | :00:00 | :00:00 | 0 | :00:00 | 0 | :00:00 | 0 | 0 | :00:00 | 0 |
| 7/28/96 | Skill 409 | 0 | :00:00 | :00:00 | 0 | :00:00 | 0 | :00:00 | 0 | 0 | :00:00 | 0 |
| 7/28/96 | Skill 410 | 0 | :00:00 | :00:00 | 0 | :00:00 | 0 | :00:00 | 0 | 0 | :00:00 | 0 |
| 7/28/96 | Skill 411 | 0 | :00:00 | :00:00 | 0 | :00:00 | 0 | :00:00 | 0 | 0 | :00:00 | 0 |
| 7/28/96 | Skill 412 | 0 | :00:00 | :00:00 | 0 | :00:00 | 0 | :00:00 | 0 | 0 | :00:00 | 0 |
| 7/28/96 | Skill 413 | 1 | :00:08 | :00:00 | 0 | :00:00 | 0 | :00:00 | 0 | 0 | :00:00 | 0 |
| 7/28/96 | Skill 414 | 0 | :00:00 | :00:00 | 0 | :00:00 | 0 | :00:00 | 0 | 0 | :00:00 | 0 |

Double Click To Run Format Grid BulG3V5pe

*Figure 3 Avaya CMS Report from version 11, dated 2002*

This visual example from 20 plus years ago represents what is still, more or less, used across many leading contact center platforms today. Very little has changed! Again, we understand why, and we appreciate that these reports may benefit your organization. But you can encounter dangerous implications if you continue with the same reports, which focus on business exclusively rather than on CX, for another ten years. "The definition of insanity is doing the same thing over and over again, expecting different results," says a widely quoted remark with no clear origin. It is critical to compare the performance of a contact center in 2002 to a similar contact center in 2024 using the same solutions and reports and to ask yourself, "Did anything truly change?"

Of course you need to measure your operational performance, but if that is all you continue to measure, your customer will always be secondary.

As we highlighted in Chapter 1, ask yourself, "Is it realistic to expect anything different if we do things the same way for another ten years?" If the answer to this question is, "We think these measurements and reports are useful, and we're certain we're doing the right thing for our customers," then you're at the right place for your operational needs. However, suppose the response is that you are not comfortable continuing with the same practices and want to consider abandoning traditional tactics in favor of an alternative approach. In that case, the following section will be of interest to you.

## How it's done!

Let's discuss why we think the above examples and other traditional contact center measures aren't good at evaluating customer satisfaction. CXs are harmed when AHT is calculated because agents typically assist clients with only one activity for which they are qualified. Agents know the AHT target for each interaction and will do everything possible to keep the contact under or equal to that level. This is frequently at the expense of not connecting with customers as empathetically as possible and not doing more to help them proactively, resulting in missed opportunities for your organization.

Another complication to agent behavior caused by strict adherence to AHT targets is increased agent-to-agent transfers and reduced First Contact Resolution (FCR). Suppose a customer's demands appear complicated, and the agent's AHT is likely to be exceeded. In that case, the agent may choose the easy route and transfer the call to another agent, essentially cherry-picking the next calls that are easier and take less time. Reports may or may not show this; regardless, the damage to the CX has occurred. Negative impacts will remain until agents are measured in alignment with CX outcomes versus AHT and FCR.

When a customer needs assistance with more than one issue, such as paying a bill, updating their address, or troubleshooting, many agents will take the easy path and transfer the call to another agent or queue. Even if the agent can handle all of those issues, they will be penalized for the call's high AHT. Most contact center technologies allow agents to enter just one call disposition, making it impossible to attribute the agent's numerous subjects of help throughout the conversation — meaning if agents are evaluated based on solving one issue per call, chances are they are going to modify their behavior to align with their compensation rather than the CX. Consequently, transfers occur and agents rush the customer off, requiring them to call in again to complete the transaction.

When reviewing contact center reports, ask yourself:

- What actions will I take, and why, how and what will I measure to ensure that the actions are appropriate?

- Will the modifications impact more than one measure, and do we have reports that capture the full scope of the situation to assess the impact of the adjustments?
- What data do you know is missing that might be used to accomplish targeted business and CX outcomes?
- Is all of the data available, but putting it together the way you want requires too much human labor?
- Do you sometimes dismiss reports because you know you won't obtain the outcomes quickly enough to justify the effort?
- Do you have a wish list of reports that have never been mentioned because they are believed to be moot, owing to IT resource limitations, data availability issues, or more broadly, a cynical attitude based on previous failures with similar endeavors?

By asking these and other difficult questions, you will appreciate what you have and utilize it, discard what isn't relevant and identify what extra is required to successfully shift the needle to the positive for improved CX delivery.

Typically, reporting is reporting for the sake of "this is how we run our business," and reporting with conventional contact center metrics alone does not drive you to focus on the CX. Take, for example, how different operations are today due to the COVID-19 pandemic, and how reporting has *not* changed one bit ...

COVID-19 shook the contact center world to its core. The influence of COVID-19 on the contact center business has been discussed in several papers and conferences. During this period, as we all know, there was an influx of home-based contact center agents, the majority of whom had never met a single peer or manager in person for a year or more. Supervisors learned to support and manage their teams without having the option of walking over and consulting with an agent who has flagged them down by hand. Yet how operations are assessed has not changed in the least, as far as our customers are concerned. Everything was okay as long as those legacy reports continued to generate the same numbers day after day!

## How to Overcome the Pitfalls: Three Strategies

It's critical to pick one of three sampling paths:

1. Conventional: Continue to use and base your CX results on conventional metrics.
2. Radical: Go all-in and toss them out the window. Then, instead, back into the cost per minute, measure something much more obscure and not linked to the length of interactions.
3. Rationalized: Settle on a rationalized method with a balance of traditional and radical metrics.

Each approach has value. Some businesses will stick with the tried-and-true method because the metrics they measure are unlikely to alter. Others will succeed by concentrating on innovation where risk-taking is encouraged, the CX is prioritized and failure is anticipated regularly to enable continuous development. However, most of us will work in organizations that must strike a careful, intentional balance that meets the requirements of both old and new viewpoints.

## Metrics of Convention

It's critical to start with where you are now with your CX strategy and service delivery methods to prepare for where you want to be. For decades, contact centers have employed the same reports based on AHT, Average Speed of Answer (ASA), Abandons and various other metrics that they deem essential to the company. As a result, the contact centers have emerged as cost-controlling centers rather than creative, customer-centric hubs focused on ensuring customer satisfaction. The measurements have historically sought to limit costs, maximize efficiency, and consequently minimize innovation. Users of standard, conventional contact center reports will argue that these are, in fact, all about the customer first and foremost; however, we beg to differ.

"What can we do to reduce the cost of serving our customers?" Every transaction handled by your contact center agent does have a cost. The number of repeat calls or chats, the total handle time, the time the customer spends in the queue and the resources required for telephony, networking, computing, call recording, quality management, workforce management and all other components needed to deliver service must all be measured for ROI. And the only sensible way to do so is to divide the sum of all the pay, perks and costs listed above into quantifiable, measured metrics. These metrics demonstrate the financial effect of not achieving or surpassing the thresholds. In the contact center, "cost is king."

The following metrics are ones that we can pretty much guarantee exist in the majority of contact centers today — thus, your Metrics of Convention.

- ASA ensures the customer is greeted and served by an agent as expeditiously as possible.

- Average Handle Time (AHT) is the average time spent per interaction talking/engaging with the customer live. It is calculated by summarizing all of the time spent handling customer interactions and divided by the number of interactions handled.
- Average Abandon Time: This metric tracks the average time customers are in IVR (Interactive Voice Response), chatbot applications or queues when they decide to not wait any longer and disconnect their interaction. Customers may abandon this for several reasons, including "Getting lost in IVR hell, trying to navigate complex menus" or "wait time in the queue is too long." Operational changes to better handle queued interactions may include adding more agents or changing agent skills. Reducing abandon rates in the IVR or chatbot could involve streamlining menus to make it simpler for customers to get what they need.
- FCR: This is the percentage of customers that get their inquiry or request answered on their first interaction. Many enterprises measure FCR by simply asking the customer if their issue was resolved and if this was the only time they had to reach out to do so. We caution against using this method alone, as customers may not fully understand the question and say yes, giving a false positive for FCR.
- Percent of Interactions Answered in Service Level: The percentage of interactions answered in a set timeframe (e.g., 60 seconds, 120 seconds) to ensure customers don't have to wait too long.
- Percent of Interactions Answered: Out of 100 interactions routed, how many were answered by an agent?
- Percent of Calls Abandoned: Out of 100 interactions placed in the queue, how many terminated their call before it was answered by an agent?

- Wrap Time: The amount of time spent post interaction for documentation, summarization and follow on actions

Legacy metrics have been, and still are, used to ensure customers are served quickly and efficiently. However, what happens when we say, "That's great! So, let's not worry about the cost and make the following changes:

- Staff to 200 percent of forecasted need, so the ASA goes to zero seconds.
- AHT is now unlimited. We want to ensure the customer gets everything they need on the first call.
- Average abandon time is no longer needed since we solved it by overstaffing.
- Percent of interactions answered in service level is no longer needed.
- Percent of calls abandoned and answered: again, it is no longer needed. Quite simply, we are going to over-staff, over-deliver and not worry about costs."

Now you must be thinking, "This is crazy. We can't staff 200 percent of the forecast!" If you think that, you will likely agree that these metrics are much more about cost-of-service-delivery than proper CX focus. WYMIWYG. "I must ensure AHT is less than 250 seconds." If this is the most important metric to a contact center manager or supervisor, somehow, in some way, decisions will be made throughout the day to ensure that metric is achieved. If it is not hit, there are likely consequences up and down the reporting chain since executives typically focus on how to make and save more money.

Not worrying about the costs of CX delivery is a non-starter for almost everyone reading this book. The reality is, nobody has an infinite budget. However, if costs were not the only decision criteria for delivering the best possible CX, we ask you to open your mind and consider alternatives.

## Radical reporting shift

*"We take most of the money we could have spent on paid advertising and instead put it back into the CX. Then we let the customers be our marketing."*
*—Tony Hsieh*

Empowering agents to do what is required to serve customer needs and do the right thing to make the customer feel comfortable and valued through the CX requires different metrics of meaning. Single CX journey rate, objective sentiment analysis, self-service rate, partial self-service rate, assisted-service rate, and customer effort score are some metrics we will discuss in this section. Let's now dismiss AHT, ASA and other conventional metrics and think about what matters to the customer. Each of the metrics in this section will more closely align with the customer than the business. The only application for the company is to report on the success of CX delivery or the need to drill deeper to see what is needed to alter the customer journey more favorably for customers.

**Single CX Journey Rate:** we created this metric to track the percentage of customers on any given CX Journey who successfully conclude their journey in a single, fluid manner and without restarting the journey within a specified timeframe. For example, if a customer needs to set up a new service: were they able to start the process, make progress within acceptable timeframes and finish the process and not have to repeat process steps? How does this tie back to the customer? This metric will give your business analysts, contact center management and executives insights into which journeys serve customers well and which ones require immediate change.

You may be asking—how does this differ from FCR? Typically, FCR is measured at the interaction level, specific to the singular channel perspective. However, in today's world, most customer journeys span multiple channels, making the FCR metric ineffective when trying to attribute "success" on the 16th contact across eight channels for a customer to *finally* achieve successful journey completion: "Awesome! This *one* phone call delivered FCR!"—"Oh, wait, the customer bounced around 15 other interactions and seven other channels to *finally* get to this phone call? So ... we did *not* achieve FCR ...? But, that is how we are *measured* for success ..." We hope you can appreciate the reason for this different metric and the need to construct methods of tracking and orchestrating journeys to line up with the journey perspective versus the interaction level alone.

How does one implement such a metric? How many of you can confidently say that you know your customer journeys and can visually map out all the points along the journey? Can you determine and report when customers make progress, stall or regress along the journey? Many things must be put in place to measure single CX journey rate.

First, it's imperative to identify who your customers are as they embark on a journey. Can you identify customers in voice and digital channels at a high rate with a high degree of confidence? If not, can you track journey progress anonymously, then associate it with the customer when they are identified at another point in the journey?

Do you have a way to store this progress and all associated context gathered along the way?

For most readers, we are willing to bet the answer to these questions is no. And, most likely, many of you are thinking, "Yeah, that is a radical departure from what we do today. It sounds hard and expensive." You may be right. But if we are focusing on the CX and not the Metrics of Convention, why settle for anything less than the ability to have your finger on the pulse of each journey, every step of the way?

Second, flagging journeys as complete will set the stage to track if the same customer starts the same journey within a specified timeframe, such as the number of days. Sometimes, this is okay! If your customers have frequent, recurring service needs, this is not a problem. However, enterprises may have a problem where customers have a typically infrequent engagement cycle. You can now see the single CX journey rate metric decrease in the percentage achieving it successfully. What has changed in your processes that may affect the outcomes? What changed in your staffing levels? Are there knowledge articles introduced and delivering incorrect information? Has customer sentiment decreased in specific journeys? Any number of items may need to be investigated. The key, however, is knowing that you need to take a deeper look and find the root cause of the impact on your CX.

Finally, establishing a method of tracking journey changes will enable rich opportunities to affect the CX. For example, if an identified customer on the website starts a transaction but then does not advance through the process further, rules may be put in place to send an SMS and let the customer know. "Hello! We see you started to pay your bill. I can help! Reply *yes*, and I will execute the transaction with the account ending in XXXX."

Why not seize the moment of opportunity and proactively engage the customer? Why not leverage the already known data, such as payment method, and let the customer respond with a simple *yes* to take care of the process? Why make the customer go through the web process or call into the IVR and talk to an agent? Any number of rules can be applied to journeys at various milestone stages. Consider all of the relevant context known throughout the customer journey.

**Remove AHT barriers:** So far, nothing discussed in this session has been too radical for most enterprises to consider. This is where it gets bumpy, so buckle up! What if you were to propose to your executive management the following:

> Let's stop measuring the performance of our contact center the way we have been for decades. What if we remove AHT barriers, commit to staffing more than traditionally needed and start to learn what it means to focus on customer outcomes and satisfaction without micromanagement of our agents?
>
> Let's commit a large portion of our marketing budget to service our customers and to measure our CX based on the length of the relationship, automated sentiment analysis of each interaction between customers and agents, and knowing that if we make this change, we are committed, from CEO to agents, to make our customers the nucleus of our business and have confidence that if we do the right things for the customer, the right things for the business will result.

Does that sound radical? We think it does, yet we believe in the recommendation. Let's take a look at some of the statistics that support the value behind considering this approach.

Bain & Company and the Harvard Business School report that "increasing customer loyalty by just 5 percent can increase profits anywhere from 25 to 95 percent. In conclusion, customer service costs should be regarded as an investment in boosting sales. Loyal customers are worth up to ten times their first purchase."

With that in mind, doubling down and pulling a significant amount of the marketing budget to focus on CX is radical but with precedent. Keeping customers happily served is a much less risky investment than focusing more spend on prospecting. Also, marketing is not immune to WYMIWYG. Marketing campaigns will target and report in a funnel manner that aligns with their metrics of meaning. Again, be disruptive and think about ways to align spending with a focus on CX.

## Rationalized approach to change

Let's walk through a scenario where we worked with a customer to implement an OKR mindset and practice.

Gina finished posting a large, sticky-backed grid paper sheet to the wall. On it was the word *objectives* and a column and three rows labeled *key results*.

"Okay," she started. "Let's talk about OKRs. Anyone know what these are?"

Nope, not a one. Everyone looked expectantly at Gina to explain.

Reading the room, she carried on convinced this was new territory for everyone.

"Objectives should have key results. For example, if our objective is to increase net promoter score (NPS), we can't simply say we plan to increase NPS. What actions will we take to raise NPS? How will we know the effectiveness of each action? This is where each objective should have key results and associated metrics to gauge performance. It's too late to wait for monthly, quarterly and annual NPS to roll in and realize you missed many opportunities to do something much sooner."

One of the customer attendees said, "I think I get it. For example, we could have an objective to implement a voice of the customer survey to learn what is working and what is not before the NPS surveys are sent out. This would give us a chance to understand what needs fixing, plan and act!"

"And there's a gold star! Excellent example." Gina smiled. "Using your example, we could have the objective as implement voice of the customer surveys with key results of highlighting what is working well and another of what is not working well. We could then have an objective to analyze these outcomes with key results of ... Actually, who wants to come up to the board and drive this conversation to completion?"

An eager project manager from the team came forward. Gina finally had to call the session to an end after 10+ sticky sheets, a couple of dead markers and a team so engrossed in the process that they failed to realize how late it was in the day.

Now that we have talked about conventional contact center management and metrics and radical/disruptive steps, the final perspective to consider is a rationalized version somewhere in the middle. As mentioned, leveraging Metrics of Convention to have a gauge on resources consumed and needed doesn't cause any harm. The harm is when it is the sole source of metrics. So meeting somewhere in the middle is a good idea. To do this, start where you are! There is plenty of middle ground to consider when you embark on a continuous improvement journey with your customers in mind at all times.

When committing to the rationalized, in-between approach, ask yourself, "What do you want as an outcome of these changes?" This is where OKRs can be established to support the corporate vision. OKRs allow for meaningful measurement of outcomes and challenge teams to establish value around everything they do and every decision made. Using OKRs transcends static, discrete metrics.

Try looking at age-old metrics through the lens of OKR

| Conventional | Objectives and Key Results (OKR) |
|---|---|
| First Contact Resolution (FCR) | Objective: Ensure all customer needs are served accurately the first time the customer engages, including transferring a customer if required.<br><br>Key Result 1: Contact center management has established customer journey tracking where interaction handling is measured. This is accomplished by properly identifying customer intent(s), accurately taking the needed action(s) to resolve and validating the right outcome(s) were attained in each customer journey.<br><br>Key Result 2: Analysis of the above-tracked journeys is automated to highlight the percentage of success.<br><br>Key Result 3: The ability to drill down to the details of interactions *not* meeting this criterion has been created to allow innovation on how to improve over time. |
| Average Handle time (AHT) | Objective: Ensure agents are empowered to serve customer needs without negatively impacting their performance review if specific interactions take longer than "average."<br><br>Key Result 1: Supervisors and contact center management have established a customer-first policy. Customer and agent sentiment analysis reporting and analytics capabilities enable contact center management to understand which types of interactions need additional time. More personalized dialogue and training are aligned to enable agents to deliver the best possible service by understanding the scenario at the beginning of the interaction. |

| | |
|---|---|
| Average Speed of Answer (ASA) | Objective: Ensure rapid, live engagement of customer interactions<br><br>Key Result 1: Orchestrate channel shift i.e., voice to chat and chat to callback to ensure the agent is engaged as rapidly as possible |
| Customer Satisfaction / Net Promoter Score | Objective: Enhance Voice of the Customer Survey Program to enable feedback in near real-time.<br><br>Key Result 1: Implementation of AI/ML to determine the customer and agent sentiment in real-time and post-interaction has enabled contact center management to know when customers are being served with Positive, Negative, or Neutral outcomes.<br><br>Key Result 2: Dashboards summarize and correlate customer and agent sentiment to allow contact center management to drill down to specific interactions, analyze for training and performance recommendations and understand where operational procedures or data access impact the outcomes and may need refinement.<br><br>Key Result 3: The quality management process has been turned upside down by evolving from hoping to review 2–3 percent of total customer interactions to programmatically reviewing 100 percent with AI/ML. This enabled the service quality assurance team members to focus on the most important 2–3 percent of interactions. |

Proof of Value per interaction as an overarching outcome can be a game-changer for any enterprise. Rather than asking customers to hold on the line for a post-call survey/email/SMS inviting them to submit feedback, an approach based on automated sentiment analysis can be employed for all or a percentage of customer interactions. The focus of the sentiment analysis is to gauge the tone and conclusion of the conversation from both the customer's and agent's perspectives.

Additionally, comprehension of the interactions can take place. Both sentiment analysis and understanding of the key phrases are powered by AI/ML solutions commonly available in the market from numerous vendors. They do not require a fleet of data scientists to implement in existing architectures. Using them can remove the need to survey your customers: you can already see their experience, whether it was neutral, good or negative, and how to assign value to the interaction. For example, did the customer achieve what they needed? Were they treated with respect, and were the proper steps taken to help them? Did the agent have the resources required, or did they have to transfer to another agent?

Each analyzed step can yield more data to continually improve customer journey orchestration, resulting in an enhanced CX. To keep the discussion in the range of rationalized versus radical, consider a starting point of tracking who, what, when, where and which journey. Whether customers are anonymous, identified or identified and authenticated, these five categories can be associated with a customer journey and tracked.

## Call to Action

Most enterprises do not have a baseline of what their CX is outside CSAT and NPS. With OKRs and more rationalized approaches to CX strategy in mind, how do you see your company establishing a baseline of where you are, and measuring progress each step forward?

- What are the metrics/OKRs that comprise your baseline measurement?
- Do you have a 360-degree view of your customer journeys?
- Do you believe you are measuring them accurately? In the right time intervals? Are they yielding the CX outcomes you truly desire or are they "keeping the lights on and costs at a minimum?" Do your current reports help your business costs or impact CX outcomes?
- If you cannot establish a baseline: How do you know what you need to fix if you don't know where you are today concerning your CX?
- Do you believe your org should consider metrics and report changes?

We covered a *lot* of content in this chapter. In the next chapter, we are going to focus on culture and provide some thoughts on how to rebuild it within your organization. This sets the stage for measuring and doing things differently.

★ ★ ★ ★ ★

# CHAPTER 4

# Data-Driven Insights and Rebuilding Culture

*"You've got to start with the CX and work back toward the technology, not the other way around."*
*—Steve Jobs*

The most significant factor standing in the way of uniting enterprises in a CX transformation is culture. Everything starts at the top: how people and divisions interact and the levels of trust and respect that may or may not exist. Your culture directly impacts your CX.

According to McKinsey research, leading CX companies will increase customer loyalty, deliver better employee experiences, increase revenue by 5–10 percent and cut expenses by 15–25 percent within two to three years. Furthermore, according to additional research, they are seven times more prone to purchase a brand-new offering and three times more likely to refer a friend or family.

If you want this outcome, you need to put CX at the center of all you do by following four steps.

- Step 1: Align common goals and set meaningful, outcome-based vision statements
- Step 2: Identify a responsible party
- Step 3: Establish a baseline on the seven key factors covered in Chapter 1:
    1. TCO/ROI take precedence over CX
    2. Corporate culture
    3. Communication
    4. No clear CX ownership
    5. Lack of innovation
    6. Technology
    7. Workflow
- Step 4: Remove barriers

## Company Culture Standing In the Way

Unfortunately, the company culture is often the biggest obstacle to implementing these steps.

The cultural issues we have experienced have varied significantly, from scenarios such as leaders hiring friends to positions of power and leaving staff intimidated to address the problems, to obvious empire building, to establishment and acceptance of a passive-aggressive approach to work, to a company that became paralyzed in making decisions for fear of upsetting anyone.

Cultural issues typically take the longest time to change. Fundamental, systemic beliefs and practices must be altered. This is why having executive acknowledgment, support and sponsorship is critical in implementing widespread change. The greatest successes we have witnessed involve executive sponsors with the ability to win the trust of and empower key evangelists to lead by example.

Even with executive sponsorship, you will always have naysayers or fence-sitters. As leaders endeavoring to effect wide cultural changes, it's essential to engage those against or in doubt of the changes to truly appreciate their concerns or questions. The phrase "It takes a village" could not be more true than in the scenario of changing corporate culture. Make sure the entire village is part of the cultural change journey.

The scenario below, which is 100 percent verbatim what we have heard in a real-world setting (name and company withheld to protect the not-so-innocent), depicts clearly the impact leadership can have on culture and morale directly affecting customer and employee experience.

> One company we worked with was absolutely leading the industry in their focus on CX. The leadership brought us in to help execute their internal transformation.
>
> The first few meetings with our sponsor showed us he understood where the pitfalls were; however, he had concerns that the rest of the team was in the "this is how we do it" mindset.
>
> "We know we can be providing a better experience to our customers; we just aren't sure of *all* the places we are falling short," Brian shared matter-of-factly. "We have done well, but we aren't renewing at the rate we should be, product is taking too long, we have teams of people not communicating or collaborating, each one seems to 'own' the customer for a certain part of our implementation and overall no one knows the full picture."
>
> "That's common," Gina started. "I can assure you, you are not alone. You stand out when you are acknowledging the issue and taking steps to fix it! Do you feel all of the leadership team shares your view and concerns?" Gina asked the question knowing it would be unlikely — however, it would give Chad and her the awareness heading into the interviews.

"Everyone is on board and so excited for the transformation. We've all talked about it and recognize this is what we need," Brian proudly shared.

The conversation with Brian continued, and we determined that to get a full grasp of the state of the union, we would approach their transformation following our established process.

- Conduct interviews with different team members from every level of the organization to get a well-rounded view of where the issues were
- Highlight the issues at an individual, team and corporate level
- Work with sponsor and leadership to identify MVP
- Introduce Agile/create scrum teams to deliver on MVP
- Establish success criteria
- Report on a monthly cadence to executives

Our goal during the interviews was to gather as much information as possible, to ask open questions and to gain an understanding of the culture, the current customer journey and where they felt they were falling short.

We began the interviews with the leadership team and realized right away the strategy wouldn't matter if we couldn't get the executive suite to shift their mindset.

"We always start initiatives and never finish," Bill stated. Bill was ultimately in charge of operationalizing the changes that were to come from this implementation. "I know where the problems are, but I'm too busy to fix them," he continued. "My team follows a process and does what I ask them to; it's the other groups that need to change so our customers are happy."

Sam, VP of sales, chimed in: "We close the sale, but then it breaks down because we have to have so many people engaged in the process. It's so hard to get the renewals because we have product issues, our follow-up is always delayed, and no one knows what is going on, and it always falls on sales."

"We continually get incomplete requirements for features, everything is always priority #1, and we are finding it hard to get releases out because even when we lock things down, the CEO will sometimes come and tell us we *have* to get this feature in. No one seems to understand the domino effect of their asks," John, the Chief Technical Officer, complained. "When I try to talk to Bill or Sam about it, they point the finger elsewhere."

When we asked the team about who owned the CX in their eyes, the answers were just as telling:

"My team owns it during the project then we are gone," Bill stated matter-of-factly. "We can't own anything once it's implemented; it's support after that."

"My team isn't commissioned on renewals enough to keep them in the forefront of their minds," Sam admitted. "I know it isn't a good excuse, but it's the truth."

"We don't even have contact with the customer. Why would we be responsible for the CX?" John asked, confused.

This pattern that emerged was a consistent one: communication was lacking and there was a culture of "not my job" when it came to who was ultimately responsible for the CX.

When we met with Brian again to share our initial findings, he shook his head. "I'm shocked that this is what the executives said. They all committed to making this work." He sighed. "We need to proceed and see how their engagement is. I'll work back channels to make sure the support is there."

With that, we set out to implement the rest of the program. Ultimately, this engagement was not one of our more successful, as Bill (the one responsible for operationalizing) stonewalled every step forward that wasn't in line with "how he would do it if he had the time." He started reassigning those dedicated to the initiative and communicated to his team that "they didn't have to do anything different from what they were doing today." With the team required to operationalize and implement the changes no longer participating at full capacity, we told Brian we needed to end the engagement.

"I see it is our culture that we need to fix now more than ever," Brian said. "I'm really disappointed in Bill's decisions."

"It's more common than you think," Chad assured him. "We see empire building all the time; it usually starts at the top."

Brian nodded. "Unfortunately, I agree. We may suffer from bright shiny object syndrome a bit too much here."

As our wrap-up conversation continued, we shared another recent, almost identical experience. "Similar to you, Brian, when our previous client had their leadership team at the table, every one of them agreed something had to change. Through the interview process, everyone had great ideas and a lot of complaints about other teams and leadership. They all pledged their support to the transformation initiative." Gina smiled. "As you can imagine, when the reality sunk in that they would be responsible for personally embodying and operationalizing the changes — in this case, they switched the focus from the transformation to technology—everyone complained that the tools used to track progress weren't accurate enough."

Brian laughed out loud. "Good to know we are in good company."

"What was different with this group," Chad shared, "is that they liberated the main culprit from the workforce (some may know this as being fired), and once the rest of the company wasn't torn with loyalty not to buy into the change, things progressed very quickly."

"Food for thought," Brian said as he glanced around the table.

"I know we discussed this at length when we started," Gina began. "Cultural change is not easy. It may mean changes to your teams and possibly even your leadership team. It takes true conviction to be a better company and change your culture to become customer-centric."

## Customer-Centric Culture

### What it is

A customer-centric culture is one in which the customer's satisfaction and experience are at the nucleus of all decisions and activities. Employees understand that customer satisfaction is their primary responsibility in the company. A true customer-centric culture begins at the top and is consistent throughout every facet of the enterprise.

One of the case studies we completed was with a large hardware and software company with more than three decades of reliability and strong customer loyalty. As the "cloud-first" strategy took hold, the company found it hard to keep pace with competitors already in the cloud. Their initial cloud versions were deployed, met limited adoption, were refactored and, in some cases, released with a lack of parity to previous versions. The need to "meet or beat" the competition was the sole focus. As a result, the CX suffered due to a lack of prioritization of what they needed most for table stakes functionality to serve their end-customer base and run their operations.

The best approach would have been for the company to establish an MVP that aligned with the desired CX. Instead, the choice to focus on the competition resulted in not delivering many of the basic needs for their customers. Whenever you do right by the customer, the right business outcomes tend to occur organically. Trying to shoehorn customers into new environments solely for your business purposes in no way guarantees delivering the right CXs. And not delivering business value and features needed will disappoint all involved.

## The challenges of becoming a customer-centric organization

### *Culture is hard to pin down in concrete terms*

How many of you work for a company that has a printed vision statement on the website, lining the walls within corporate facilities, and a set of leadership principles meant to ensure all employees are aligned on the company's mission? And how many of you have developed cynical thoughts of these based on living daily in an alternate reality where actions and outcomes do not match these statements?

Vision statements and policies/procedures don't create a culture; communication, collaboration and empathy do. Like you, we could quote 20 cliché phrases about culture. These statements can be very motivational; however, the only thing that matters is what your organization practices on a daily basis.

Culture should not be imposed or enforced on employees from the top; instead, it should be nurtured by empowering them. When leadership sets a positive example from the top down, culture can be embraced authentically. Otherwise, well-intentioned phrases, vision statements, and leadership principles can be misused as weapons.

Many of us have likely experienced or witnessed situations in our careers where a boss criticizes an employee, like in the case of Joe, who had already put in extra effort but was still asked to take on non-critical tasks, undermining his dedication to customer satisfaction.

Showing our age here, a clip from the classic movie *Office Space* comes to mind where Jennifer Anniston is asked to express herself with a mandatory amount of pins on her suspenders. You can't "force" personal behavior; you must encourage it and build the trust of teams to *want* to align themselves with the culture demonstrated by true leaders.

**"You Know What? Yeah, I Do. I Do Want To Express Myself, Okay? And I Don't Need Thirty-Seven Pieces Of Flair To Do It!"**

*Figure 4 Office Space, directed by Mike Judge (1999; Judgemental Films, distributed by 20th Century Fox)*

## *Empowering the workforce accordingly*

People — not directives, buildings or systems — are the core of any culture.

One of our customers set the gold standard for empowering their workforce. Leadership set the vision using OKRs (see last chapter), then empowered their teams to reach those objectives. One team in particular spent some time talking about what the OKRs meant to them. Because the culture was so empowering, they were allowed to determine how they would achieve the objectives without being *told* what to do. This grew into an increased commitment to the company by the employees and, ultimately, their customers. The broader impact was the bar being set for other teams to adopt. When you see individual contributors living and breathing the culture of their own volition, you know you are on the right track to delivering on the promised mission and vision statements for your customers. High-functioning teams leverage company tenets and leadership principles effortlessly.

## *Perception and reality*

How many of you have worked at a company with a perception of invincibility? Of being on top of the market in your space?

Most companies we have worked with have an inflated view of either their CSAT or their employee satisfaction scores, and sometimes both!

> We were midway through our first week at a customer site when Katherine, the VP, proudly announced, "Our CSAT isn't an issue; we have improved four points in one year!"
>
> "That's fantastic! What do you think contributed to this increase? What did you change with your delivery to have grown so much so quickly?" Chad asked curiously.
>
> We watched a lot of side-eye exchanges and a few quiet murmurs occur around the boardroom table.

"Why don't we take fifteen and pick this up where we left off?" Gina asked the team and was met with the agreement of the group.

As we were preparing the room for the next session, one of the supervisors, Angie, hung back from the rest of the group and told us, "Last year when CSAT became a major part of the compensation plan, we started asking our customers to rate us a ten so that our pay wouldn't be impacted." Angie's face turned red as she shared the story. "I probably shouldn't have told you that — I think everyone knows that is what is happening, but no one really talks about it."

Chad and Gina shared a knowing look; yet another great example of WYMIWYG.

Everyone started filing back into the room, and Angie took her seat with the rest of the team.

"There are a couple of stats we want to dig a little deeper into," Chad started. "How did the total number of surveys completed month over month and year over year compare?"

Gina added, "Can you also share with us how much your customer churn rate improved with your CSAT improvement?"

John started typing away on his keyboard — he was projecting numbers up on the shared screen. "Well ..." he paused; everyone in the room was silent. The numbers were right there for everyone to see. "Our churn rate has actually increased and the number of surveys has decreased."

Silence filled the room again. Katherine was in shock. "We never moved forward with changing any of our processes because we saw the improvement that we needed."

The false sense of security this customer had when they saw the increase in CSAT led them to lose their edge and sense of urgency concerning pleasing their customers.

It is wise to ensure you have reliable methods of gauging where you are with respect to all your OKRs, not just those that are tied to revenue. Assumptions such as "all must be good now that CSAT is improving" are a good recipe for being in a perceived position of greatness versus an established, reliable quantification of how well you are doing for your customers.

## *Roadmap*

Starting with the end customer in mind may disrupt your approach to your business roadmap. We've seen many roadmaps for large enterprises, and they have focused on everything from systems to sales — basically anything business-related versus an emphasis on the net CX.

In one of our engagements, our customer was showing us the personas that they used for the focus of their product development. Shockingly, they did not include the end customer. When we mentioned this during the leadership review of the roadmap for the upcoming quarter, one leader responded with, "We just need to focus on what they will buy." We had an open dialogue with the team about the importance of selling and asked if they were able to ensure business value with their proposals to customers. We discussed the CX of being told what the new products were versus ensuring we were solving a pain point. This same leader was the one who continuously prioritized their personal preferences on the roadmap instead of weighing the requests from the end customers. A few of those features had multiple customers requesting the same functionality for over a year.

Obviously, we took issue with this conclusion. We tirelessly worked to win all stakeholders over to the mindset of putting the end customer first, highlighting the need for the direct customers to be the nucleus of the roadmap deliverables. This change in direction on the roadmap was immediately met with improved CSAT scores the following quarter — the customers appreciated being heard!

Another of our customers in a very similar situation, unfortunately stayed the path and ended up paying for their decisions. That situation was a little more heated, as you will note in the exchanges below.

> Jennifer blurted, out of exasperation at not having her perspective valued, “Look, Sam, I don’t think we are using the data we have from customer feedback. You keep saying ‘Customers will never want that. Here is what they really will get excited about,’ yet all of our surveys, focus groups and direct feedback to our agents state quite clearly that their needs and wants do not align with your personal vision.”
>
> Sam’s face reddened, but he remained calm.
>
> He then said, “Jennifer, I hear you, and you are right. I just think that if we take my path, we will get even better results.” His tone was clearly patronizing, and it was obvious he had no such thoughts about Jennifer being right.
>
> Jennifer replied calmly, coolly, but directly, “Sam, you still are not listening. We need to orient toward the customers’ vision and form our strategy out of it. We are doing great in the market right now! Focusing on shiny objects and new technology couldn’t mean any less to our customers. We need to stay the course, prioritize based on the feedback and data and then set our product roadmap!”

In the end, Sam made the executive decision to define the roadmap to deliver what he felt was most important and valuable. Interestingly enough, Sam was not there one year later as sales and customer satisfaction dropped significantly. Customers found the services and products they wanted elsewhere, and Sam's shiny objects were never attained.

From this story, you can see your roadmap should include CX as a key objective. For some businesses, this new focus on the CX is a significant step forward. For others, it may be a light uplift to their current trajectory. In either case, the roadmap should orient on laying the foundation for guiding principles in every facet of CX within your organization.

Think big with your roadmap! Aiming too low is an easy way to miss opportunities.

We've discovered several important questions that we ask our clients to help them identify their goals:

- What is the corporation's appetite for change?
- What data or tools do you have to measure where you are today and where you believe you need to be?
- What is the difference between what customers desire and what they get?
- How realistic is it to set three-month, six-month or nine-month goals?
- Are you committed to the objective of putting CX at the nucleus of your business?
- How can you achieve a competitive edge in terms of CX?
- What competing objectives are splitting the team's attention, and how do you plan to mitigate that going forward?
- Now that everyone is on the same bus, is everyone in the right seat? Are you utilizing your team's talents to their utmost capabilities?

## Case study

A great CX strategy begins with a goal focused on what customers care about and empowers frontline employees to deliver. Big airports may be the most demanding CX locations. The McKinsey Study of an American airport's efforts to create a unique customer experience exemplifies how such a shift might occur. As you read the summary below, look for how this can apply to your place of work.

1. Transformation Team and Initial Analysis: The leadership assembled a large team to redefine the airport's objectives, focusing on passenger comfort and efficiency. A multilayer diagnosis, including complex analytics and passenger tracking, helped create a profile of passenger satisfaction and dissatisfaction, leading to a new customer experience program.
2. Understanding Passenger Needs and Emotions: The airport used Disney's Compass Model to identify different kinds of passengers and their emotional mindsets, needs, wants, stereotypes and emotions. This helped in planning a customer satisfaction program tailored to the real needs of the passengers.
3. Employee Engagement and Experience: The airport's leadership realized that frontline employees were crucial to delivering a positive CX and employee experience mattered more than they had previously considered. When they saw that employees lacked practical knowledge and training, they realized everyone needed to be on board for success. They addressed concerns like staff shuttle delays and lack of proper break rooms to improve employee satisfaction.
4. Collaborative Leadership and Common Goal Setting: The airport's comprehensive leadership team, including all 60 or-

ganizations, worked together to decide on a common goal of delighting and appreciating each passenger. This collaboration fostered a culture of unity and shared responsibility.

5. Cultural Practices and Behavioral Changes: New cultural practices were fostered, emphasizing awareness, proper communication, eye contact, helpfulness and interdepartmental camaraderie. These practices served as a connecting point between employees and the underlying CX objectives, simplifying what was required for everyone within the airport.
6. Innovation and Continuous Improvement: Innovation teams were created to generate impactful concepts around comfort, safety, speed and convenience. Over several years after the initial implementation, a group of over 50 staff members met regularly. The purpose was to strengthen and enhance a common sense of purpose by addressing challenges and fostering creative advancements.
7. Results and Impact on Culture: The changes led to increased retail expenditure per passenger, improved stakeholder participation and enhanced interdepartmental relationships. *The CEO and his team's commitment to spending time on the front lines and focusing on customer experience versus metrics fostered a culture of empathy, collaboration and customer-centricity.* In conclusion, the case study highlights the importance of understanding both customer and employee needs, fostering a collaborative and empathetic culture and committing to continuous improvement and innovation. The cultural shift not only improved the customer experience but also enhanced employee satisfaction and interdepartmental relationships, leading to tangible business results.

## Establishing the CX Strategy

Transitioning from this study to a broader context, we often encounter a fundamental question: How do we initiate such a transformation? The answer lies in the identification of a dedicated owner for the enterprise-wide CX strategy—a crucial step that many organizations struggle with.

All too often, believe it or not, we see that nobody is officially assigned or empowered to control the CX strategy. We walk the halls of businesses and see the Mission Statements painted on glass, large posters, or other media formats. Still, when we ask, "Who is in charge of the CX that aligns with your mission statement?" we either hear, "We don't know" or "We don't have any one individual in that capacity ..." If this is true for your organization, you aren't alone, and we understand!

Within the last five years, more enterprises have established CX Officers (CXO) or comparable positions to establish and drive this facet of the business. In our experience, though, these roles have not been fully staffed or funded to effect change. In turn, they are largely unable to achieve their strategy due to individual business units taking independent actions contrary to the CX charter from the CXO.

In other cases, companies often assign the responsibility to the vice president/senior vice president of customer service. Our observation is that this places the focus of the company's lifeblood within the team that measures success in terms of Metrics of Convention, as discussed in the previous chapter. This puts the emphasis on cost containment versus driving CX innovation.

We have also witnessed situations in which the C-levels hadn't bought into the need to change. These roles were often lumped with the revenue and growth functions, leading one to believe it is more about acquisition than the experience.

Once you identify who *owns* the CX, the next step, as outlined in the case study, is to figure out who *knows* what the CX should be. Before thinking about your tactical issues — ask, "Am I genuinely empathetic to the end customers' experience?"

## Ask the customers

When was the last time you asked customers how they wanted to contact you? What is their channel of choice? How often do they want to hear from you, and what do they want you to ask them? What matters to them? What would make them want to work with you more?

We experience an "oh shit" moment with our clients when we sit down to discuss "what the experience should be." We often hear "I hate being on hold" or "I wouldn't want to be transferred," and usually 10–15 minutes into these conversations, we realize that they haven't mentioned the actual customer yet. Knowing how to strike that balance between anticipating your customers' needs and projecting your thoughts/experiences into the process is imperative. However, you always need to talk to your customers — you don't always know what is best! If you can't get past yourself, you can't get to your customer.

### *Steps to determining the CX experience*

**Step 1:** Define the needed end-state CX, which may include researching industry best practices, assessing rivals' CX tactics and interviewing customers concerning their expectations and requirements.

**Step 2:** Perform a current-state evaluation to determine the organization's existing maturity and capabilities in terms of CX.

To do this, look at the following four areas to analyze the capabilities of an organization, and rank each one according to its maturity level:

- Core competencies that are below industry norms.
- Complying with industry standards in selecting key competencies.
- Adherence to industry norms in terms of fundamental competencies.
- Core capabilities that exceed industry norms.

**Step 3:** Compare the desired end state to what you have today. This helps you spot any gaps that may prevent you from achieving your optimal state. So you need to:

- Determine what needs to be done to close the gaps.
- Define the particular efforts required to address these discrepancies in detail.
- Identify resource requirements, staff accordingly and implement the actions necessary to fill these gaps based on the findings of this evaluation.

**Step 4.** Create a blueprint. The outcome from the previous step establishes a blueprint for your CX strategy.

Blueprints must be reviewed at frequent intervals. For instance, when the corporate climate has changed, you may need to make changes to the blueprint. Or results from the customer feedback loop may also necessitate reviewing the blueprints. Every bit of feedback offers a chance to get better. Organizations should have a best practice to respond to both good and negative consumer feedback. Positive feedback may suggest an activity that calls for repetition and implementation on a bigger scale, while negative feedback typically shows areas that should be corrected.

Overall, the data can show where changes are required and where your company has opportunities to shine.

So you need to utilize the data and disseminate it internally. Consider how you can improve, and, above all, do something about it. It's pointless to gather input, evaluate it and devise a strategy if you don't follow through. We have seen many customers go through the motion of gathering customer feedback, having an OKR related to a percent of surveys responded to and achieving that OKR — then doing absolutely nothing with the customer feedback. One client, in particular, started getting direct feedback from their customers saying "Don't ask me to fill in your surveys anymore if you refuse to do anything with the information we have previously provided."

As noted earlier — measuring at the wrong frequency is a pitfall we want to help you avoid. Customer feedback should be responded to within days, and action plans to address the issues, based on severity, should take weeks to deliver, not months. When you are undergoing your transformation, we recommend that your implementation team meet weekly and the executive steering committee monthly.

## Call to Action

- Returning to the example of a "clean" petri dish: what do you want in it? What would you exclude? What would the nirvana outcome be? Why? What differences can you imagine based on a clean culture from scratch?
- Does your organization have an aligned vision with a focus on CX?
- Does your organization have a dedicated CX owner?
- What cultural behaviors are accepted at your organization that stand in the way of providing exceptional CXs?
- Does every aspect of your business orient toward your North Star?
- How willing is your organization to change? Is the objective to radically alter the consumer experience or merely to improve it on the margins?
- What is the difference between what customers desire and what they get?

★ ★ ★ ★ ★

# CHAPTER 5

# CX as a Differentiator

*"Building a good CX does not happen by accident. It happens by design."*
*—Clare Muscutt*

This chapter will walk through how to design your Customer Journey Map including the process for implementation that we have touched on within the book and have found successful in corporations ready for change. When your culture fosters innovation, you will find a way to succeed.

Rewiring a firm to offer an outstanding CX is a lengthy process that takes from months to years, and necessitates high levels of participation from both corporate executives and frontline employees.

We encourage a three-step approach:

1. Focus on the foundation: To improve your customers' experience, you need to identify the very basic steps you can take first. This may be something like shortening the onboarding process, decreasing attrition or creating processes to improve consistent experiences for your customers.

2. Build the bridge/elevate the experience: This step focuses on building the bridge to the desired end state. You've addressed the basics. What is the next level that you need to optimize to truly begin demonstrating to your customers that you are operating with them in the forefront of your mind?
3. CX Nirvana: Your ideal end state will no doubt include long-term customer relationships, residual revenue, and a preoccupation with your customers' desires. Remember the increased ROI and the improved CX and EX are concrete benefits that will automatically be derived from this experience.

## Focus on the Foundation

So, where do you begin? We encourage evaluating where you are within your CX life cycle and where you want to be. Only then can you determine the right people with whom to work and why and how. As Steve Jobs brilliantly stated, "You got to start with the CX and work backward." Understanding how to provide clients with the experience they require will be aided by mapping your customer journey. We use workflows to sketch out the actions and procedures that will create the CX you envision.

Once you've finished, you'll have a visible Customer Journey Map and a graphical workflow to help you understand your client's needs, what their interactions with you look like today, what you want them to look like in the future and how to get there.

The Steps to Implementing a CX Transformation:

1. Define what success and the Customer Journey aspirations are
2. Assess the current situation through one-on-one interviews, identifying both successes and gaps between reality and nirvana

3. Identify all the teams within your company that interact with the customer from marketing to renewal
4. Capture each step of the workflows for those teams, including inputs, outputs, and cross-functional dependencies
5. Run business simulations to facilitate cross-departmental cooperation, increase empathy and elevate the employee experience
6. This is where we break down the identified gaps into actionable work —
   - We recommend starting with the MVP and anything that could be perceived as low-hanging fruit.
   - Next, dive into the issues that may take a bit longer to resolve that elevate the experience.
   - The final group of issues addressed falls into the category of long-term, strategic initiatives which, when resolved, will provide an industry-leading level of CX.
7. Establish accountable parties for each workstream
8. Create and focus teams to execute the vision

To help illustrate this process, We will build on the example from the last chapter with the company that had a "bad apple" executive blocking the change. Throughout the story, we will note steps 1–8 in brackets. Several months after completing our initial engagement, we were re-engaged by the company. The executive that was blocking the change had been let go, and both the new and existing members of the team could see the writing on the wall that if they kept going in the same direction, they would not be around for much longer.

After a few initial sessions, we realized that not much had changed. The same issues we initially identified hadn't resolved themselves, and in some cases, the gaps had increased.

"We don't believe we need to do the interview process again, Brian," (2 of 8) Chad said. "After speaking with a few of your team members over coffee this morning, it appears the issues remain the same, possibly with a few new fractures. Has your CX mission evolved to the extent that the leadership team is now in agreement?" (1 of 8)

Brian smiled. "Happy to say at this time, we all vehemently agree on the mission."

"With that being the case, I believe we have the gaps identified from our last interaction from where we are today to what the vision is," Chad carried on. "We believe it's time to review the customer journey from engagement to renewal as a comprehensive team." (3 & 4 of 8)

Gina was at the whiteboard documenting the phases of the customer journey:

- Initial Engagement/Awareness: Marketing, Business Development
- Purchase/Buy: Sales, Engineering
- Deployment/Implementation: Professional Services, Engineering
- Optimize and Expand: Customer Success Management/ Sales/Engineering
- Renewal: Customer Success Management

"Brian, these are the phases of the customer journey we would recommend for a company like yours. Are there any steps or teams that are not represented?" Gina asked.

"Maybe our support team," Brian said. "I guess they are part of deployment, optimization and renewal because if they aren't responsive, it will make everyone else's jobs harder."

"Fantastic insight!" Gina exclaimed as she updated the whiteboard.

After coming to an agreement on the five stages of the Customer Journey within the company, the next few weeks were spent outlining every customer touchpoint under each team's purview. This included documenting internal processes that supported each interaction and the contributions from each of the cross-functional teams.

When cross-functional teams are at the table, it's often challenging to ensure everyone aligns with the common goal. While the simulations are an incredible tool to bring everyone together, this customer wanted to know before starting just how far apart his teams were.

We decided to use how they were measuring their customer health to illustrate the importance of bringing the team together through the simulation (5 of 8). Everyone was gathered in the boardroom with representation from each of the different groups within the company.

In the following dialogue, we think you will gain an appreciation for how different the views of a customer's overall health and satisfaction can be within a company. This is exactly the main point of this book: if everyone isn't focused on the net CX as the core objective, your company will invariably see a divergence of perspectives and status.

"Team, we are here today to get a sense of the overall health of your sample customer A," Chad began. "We would like each of you to walk through where you see the health or status of the engagement with this customer."

"I'll go first." Bill from sales stood up. "We closed the sale on this engagement three months ago, the sales process went smoothly. I'm waiting for phase one to be deployed, which is running late from my understanding, so we can go in and sell the next business unit. For me, this account is in yellow because I don't have any proof of our success to build on the next sale."

"Hold on," James said. "We are only late because product missed a release date. Our professional services team is on schedule with what we can deliver. However, as a result of the missed release, we are in red, but this is not our fault."

Patak from product shook his head. "We had other priorities, and it was far more complex than we anticipated, so I see this customer as yellow."

"How can you say that?" James shouted. "You don't even speak to them to know how angry they are. And part of that reason is that we didn't know you would miss the release until after it was supposed to be available."

"Let's table that discussion for a moment," Chad interjected, trying to give the room time to cool down. "Sarah, where does the speech team have the customer health rated?"

"Our portion is on time and hitting our targets. The customer is definitely in green!" she declared with extreme confidence.

Gina stood at the whiteboard documenting each group and their perception of the customer's health. "I think the only group left to weigh in is the customer success team. Carol, where do you see the health of the customer?"

"Well, when I spoke with the customer last week, they weren't happy. They haven't threatened to leave yet, so I guess yellow?" she timidly replied with total uncertainty.

"When I talk to the customer, they are happy with what we did!" Sarah shot back.

Drawing a deep breath, Gina began, "Team, let's take a moment to summarize what we just heard. What we have observed is a singular focus, group by group, on your perceptions, not the net CX and outcomes. Each of you is focused on blockers and impacts on your individual teams, instead of coming together to ensure the CX is united across your organization.

"For example, Bill, you seem to be more focused on the next sale, more so than the successful delivery of the first engagement."

Shifting her gaze, Gina addressed each person. "James, we see your hands are tied due to the missed product release. You were not able to manage the customer's expectations because you were not notified that the release would be late.

"Patak, you don't seem to have an awareness of the impact the missed release is having on the customer's overall health.

"Sarah, you are only taking into account the work your team is doing for the customer, not the overall project.

"Lastly, Carol, you perceive status based on whether a customer will leave or not versus the actual experience they are having."

The team was quiet. Gina calmly continued, "Does everyone see how the different focuses for each group are impacting the overall experience with your company for this customer?" She paused for a moment and continued.

"With one green, three yellow and one red, what would you say the overall health of this customer is?" Slowly looking around the room for effect, Gina continued, "Which of you owns the CX to make a final status declaration and own the ultimate outcome for the customer?"

There was no reply, which was not a surprise since the interviews had indicated this to be the case. However, this was perhaps the first time the team could see how it directly impacted their customers.

Chad broke the silence. "Green is probably the easiest place to start. What criteria being met would constitute a green status?"

The team started offering up measures, while Gina documented their input for everyone to see. We continued going through all the groups until we had a full list of criteria that held a customer in green, yellow and red status respectively.

"Looking at the list on the board, can everyone agree with the criteria laid out here?" Chad asked. We looked around the room to a group of heads nodding.

"Looking at our customers holistically will really help us know who needs account plans and who doesn't," Carol shared.

"It will help us plan what goes into each release a bit better as well," Patak admitted.

As we wrapped that session, our sponsor smiled and said, "Point proven, I had no idea we were so siloed."

"We want to assure you," Gina replied, "you are in great company. The best part is you are aware, and now we can action it!"

The next eight days were spent running each team member in the company through the business simulation. The feedback and experience were very similar to the examples we've highlighted throughout the book. While we had some initial naysayers, in the end, there were far more "aha" moments, and it was evident it increased the morale of the teams as well as the appreciation for what everyone deals with in their day-to-day responsibilities.

In parallel, we identified the MVP (6 of 8) from the existing list of issues and within that list, the low-hanging fruit, we acknowledged that larger initiatives would be required. As the cross-functional dependency conversations occurred, we could all sense the different collaborative approach versus the finger-pointing and defensiveness we initially witnessed during the customer health discussion.

We then created several scrum teams (7 & 8 of 8) to begin to address the MVP, and within two to three weeks, the teams started seeing and experiencing the impact of the small changes they were making.

## Build the Bridge/Elevate the Experience

In the journey to elevate the customer experience, you need to optimize your processes to lay the groundwork for success. To achieve this, you can adopt a strategic approach that involves eliminating, automating and simplifying tasks that hinder efficiency. By identifying and removing unnecessary complexities, you can streamline your operations, enhance productivity and create a more seamless experience for both customers and employees.

Standardization is a key component in process optimization. Establishing standardized workflows and procedures ensures consistency in service delivery, minimizing errors and increasing overall efficiency. This not only improves the customer experience but also empowers employees to work more cohesively and efficiently, leading to better outcomes across the board.

At the heart of successful process optimization lies the commitment to customer-centric behaviors. Placing the customer at the core of every decision and interaction enables you to gain deeper insights into their needs, preferences and pain points. Armed with this knowledge, you can tailor your offerings, communications and support to deliver a personalized and meaningful experience that resonates with your customers on a profound level.

Reducing Time to Value (TTV) and Time to Resolution (TTR) is another vital aspect of elevating the customer experience. In today's fast-paced world, your customers expect swift responses and resolutions to their inquiries or issues. By shortening TTV and TTR, you can meet these expectations, demonstrate responsiveness and build a reputation for reliability and efficiency.

## CX Nirvana

Achieving CX Nirvana is the ultimate goal for anyone looking to create an exceptional CX that sets them apart from the competition. To achieve CX Nirvana, you need to exceed customer expectations. Your customers will remember extraordinary experiences that go above and beyond the ordinary.

At the core of this journey lies customer advocacy, where satisfied customers become passionate advocates for the brand, driving organic growth through positive word-of-mouth referrals and testimonials. When customers become advocates, they not only remain loyal but also actively promote the brand, acting as valuable ambassadors in the market.

To achieve competitive separation, you need to go beyond meeting expectations and embrace customer obsession. This means understanding customer needs, preferences and pain points on a deep level and proactively anticipating their future requirements. By staying one step ahead and offering innovative solutions and personalized experiences, you can create a unique value proposition that sets you apart from your competitors and positions you as the go-to choice in the market.

A natural consequence of achieving CX Nirvana is rapid sales growth. When customers are delighted with their experience, they are more likely to make repeat purchases and, in turn, become repeat customers. This cycle of loyalty and advocacy creates a powerful positive feedback loop that drives sustainable growth and success.

## The Mechanics

The basis for the implementation aspect of our CX transformation framework we developed is based on the Deming Cycle. This cycle may be used in any scenario where improvement is necessary, and its foundation is in the Plan-Do-Check-Act cycle.

- Plan: Design or revise business process components to improve results
- Do: Implement the plan and measure its performance
- Check: Assess the measurements and report the results to decision-makers
- Act: Decide on changes needed to improve the process

### Plan

It's critical to have all pertinent teams and organizations united and on the same page while approaching an identified gap for remediation.

Planning tools can vary from spreadsheets, standard project management tools or specific project management software. The only thing we caution about here is not to let the tool become the focus. For one of our customers, the fear of change showed up in so many different forms that they internally derailed the project's progress for almost three weeks as they debated on which tool would best show the project's progress. The ironic part of the entire situation was they stopped any progress on the project for almost a month to debate how to show progress. We believe in progress over perfection — get to the "do" part as soon as possible.

## Do

The actual implementation of the project is the second step. This is a simple stage if you've ensured that everyone in the company (from frontline employees to senior management) knows the project's logic, reasoning and expected advantages.

Budgets and resource allocation for execution may be hard to gain if you don't have that buy-in, so it's best to include important stakeholders early and communicate regularly. Ensure that team members are on board with any frontline changes and that critical milestones are clearly defined during the delivery cycle. Allow your team members to analyze and break down the solution so you know its true limits before it is deployed in the field. Failing fast and often is encouraged here.

There are some basic implementation strategies most successful companies follow. Whether Agile, Waterfall, or "Wagile," the most consistent factor is that they all experience greater volumes of success when they deploy the concept of visible work.

Visible work is critical not only as a contributing factor for a successful stand-up meeting but also to the results of your overall program. Visibility promotes transparency of status, from team members to the entire enterprise. While having your workflow visible is a new concept for most companies, this will help eliminate critical information from hiding in individual and team-level silos, cutting down on the time your team will spend managing email and noise from other teams and preventing any tasks from being overlooked. One of the most important parts of planning sessions is seeing, understanding, discussing and committing to cross-dependencies between teams. Without a visual way to express all that is required, it is difficult to ensure everyone knows what must happen and when and what the impact will be if not achieved.

But you can't just display it, you need to think about the format and how you show day-to-day progress in an easily consumable manner. We've had customers who wanted to use a traditional project plan with the tasks listed out (think Gantt chart). This is rarely as useful to those not updating it daily. When was the last time you saw anyone outside of project management pull a project plan for a status overview?

For those of you who are used to working with a product owner, one of the first things we hear from the role is how having work visible helps them understand the impacts of their demands. We've all been there when leadership throws more critical "stop the presses" pieces over the wall and makes everything a top priority. If everything is a top priority, nothing is. A revelation such as this can have a long-term impact on improving your employee experience as a level of respect is gained for the throughput the team is producing.

Additional benefits of having your work visible include allowing the team to efficiently combat the top time thieves that prevent you from getting the job done, including unplanned work, too much work in progress, conflicting priorities and dependencies — on other tasks, resources and technical debt.

## Check — monitor the activity

When the project begins, you must establish a baseline of the current state of affairs: where you are today and what the end goal looks like. Then make sure you have chosen well-defined metrics based on the customers' requirements and, thus, their experience in mind. With all this in place, you can then establish appropriate milestones to measure along the way. We have found the most effective time frame to "check" or measure is intervals no greater than six weeks.

Regular monitoring necessitates a grasp of the initial state and the fresh state to compare the two. Assessments of the difference between these two or more states must be taken frequently, as services might alter depending on various factors such as the season, holidays, sales volumes, etc.

Checking isn't just about metrics; it's also about talking to your customers and gathering feedback.

Some projects like software and hardware deployments, where the products are well-defined, are easier to monitor than experience projects that can be a little more difficult to scope and measure progress adequately. As a result, it's critical for the CX manager to keep an eye on the new project's release to verify that it properly meets the company's goals and customer expectations.

## Act — improve the project

Since plan, do, check, and act is a perpetual cycle, rather than a past iteration, you should be continually acting on your customer's feedback and constantly evolving your processes.

To keep exceeding client expectations, you need to brainstorm new ideas to improve and correct your current processes and tools. As previously said, each of these projects can be tiny; nevertheless, the cumulative effect of numerous small modifications can be rather amazing!

## Brian's Success

Over the next eight months the teams continued to work together, the MVP was addressed by month three, and elevating the CX became the focus for the entire company. Many of those initiatives took in-depth planning, product releases and some significant process changes. What we found, as the months passed and the progress slowed, was that the team maintained the vision. At the beginning of every project call, the CX mission statement was stated so the team never lost sight of their direction. When we completed the employee checkpoint in month eight, the issues, gaps and responses were significantly improved, and not surprisingly, the CSAT mirrored the improved employee satisfaction. Departments were working better together and the company was growing. The improvement to the bottom line was also evident.

As part of the transformation plan, we revisited the customer every quarter, running simulations for any new team members that had joined, and if we saw things slipping back into old patterns, we ran coaching sessions to help the team refocus. Some of our customers hire Agile coaches full-time; some choose to keep it in a more consultative setup. The important thing to note here is that this, like any learning, is an ongoing process and not a one-and-done kind of engagement.

Over the next three years, as Brian and his team worked toward their CX Nirvana, we would visit them regularly. As the CSAT grew to the level they had first envisioned, the ROI grew in parallel. We sat down with Brian and started the discussion about continuous improvement to perpetually enhance their CX, maintain happy clientele and boost loyalty. This changed ongoing engagements to focus on innovation rather than operations.

## Building Effective Workflows to Enhance Your Customer Experience Journey

Depending on your customer journey, you can develop one of five workflows to substantially improve the CX with your organization.

- Initial Engagement/Awareness: Marketing, Business Development
- Purchase/Buy: Sales, Engineering
- Deployment/Implementation: Professional Services, Engineering
- Optimize and Expand: Customer Success Management/Sales/Engineering
- Renewal: Customer Success Management

**1. Initial Engagement/Awareness workflow**

The awareness stage of the customer journey is all about delivering information relevant to how your product/services can help your customers overcome a common pain point and assist them in solving their problem by selecting the proper solution.

Designing a workflow around this process will allow you to understand where clients have their initial engagement with your brand, what you wish for them to do next, and the various actions required to make it happen. Often this is led by marketing or your lead generation group.

**2. Purchase/Buy workflow**

Once prospective customers comprehend your product's advantages and know-how, you'll want to focus on the sales cycle. In designing this workflow, you will often include a sales playbook. When transactions aren't made, the buy workflow might provide choices to lead customers back to the point of a possible purchase.

**3. Deployment/Implementation workflow**

Once your customer has decided to engage with your company, a deployment procedure flow will ensure you deliver the best onboarding experience possible. This flow will also allow you to organize your internal delivery teams and provide a seamless customer experience.

**4. Optimize and Expand workflow**

Construct a procedure for your product upgrades, including how and when to communicate the information to your customers as part of the customer journey to increase the level of trust. Letting customers know that you are constantly upgrading your service delivery will facilitate their use and build their affinity with you.

**5. Renewal/retention workflow**

Customer journey mapping and tracking is a continuous improvement effort. Innovating new ways to engage your customers is an opportunity to keep your finger on the pulse of the CX.

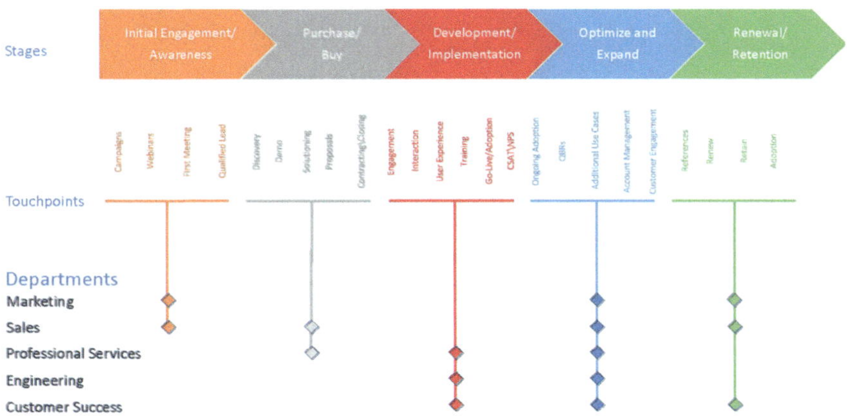


*Figure 5 Workflows*

## Call to Action

- Start where you are
- Book your business simulation
- Map your customer journey
- Align what you are measuring to your CX Goals
- Implement
- Continuously improve
- Enjoy the journey!

# EPILOGUE

We want to thank you for reading our book! Please provide feedback and reviews wherever you purchased this book.

## Summary: Embracing the future of CX

As we conclude our journey through the chapters that have unfolded the importance of CX and its transformational power, we find ourselves standing at the precipice of a new era. The lessons learned, insights gained and strategies explored have all led us to a profound realization — the future of CX holds both immense challenges and boundless opportunities. Overall, these chapters provide a comprehensive foundation for understanding and implementing your CX transformation, covering topics such as CX assessment, leadership, data-driven insights and customer journey mapping.

**Chapter 1: The Foundations of CX**

In this chapter, we delve into the foundations of CX and its impact on businesses. We witness how customer expectations have evolved, demanding personalized, seamless and memorable experiences across all channels. We learn that CX is not a mere add-on but a strategic imperative that drives growth, loyalty and differentiation in today's competitive landscape.

The chapter further discusses the seven key factors that we believe impact overall CX. These factors include:

1. TCO/ROI taking precedence over CX
2. Corporate culture
3. Communication
4. Lack of clear CX ownership
5. Lack of innovation
6. Technology
7. Workflow

Overall, the chapter emphasizes the need for organizations to prioritize CX and align their metrics, culture, communication and ownership accordingly. It highlights common challenges faced by companies and provides insights into addressing these challenges to enhance the CX.

**Chapter 2: Using Business Simulations to Effect Change.**

The chapter covers the concept of business simulations, including the focus on communication, financial analysis, strategic thinking, teamwork and communication. It highlights the benefits of simulations, such as experiential learning, skill development in a realistic setting, enhanced collaboration and increased participation. Successful simulations can align employees with strategic goals, address communication and collaboration breakdowns and drive culture change within organizations.

Take a moment to reflect on the potential benefits that business simulations can bring to your organization. Identify particular challenges that simulations can effectively address and recognize the educational outcomes they offer. These outcomes encompass the ability to instill customer-centric strategies, plan with a clear objective, take ownership of results, encourage a culture of constructive feedback, and nurture continuous learning and growth.

**Chapter 3: The Human Element of CX**

In this chapter, we emphasize the enduring importance of the human element in CX. Despite technological advancements, the human touch remains a vital aspect of delivering exceptional experiences. While automation and self-service are valuable, they must be balanced with empathy, emotional intelligence and genuine human connection. Organizations that prioritize human-centric design and foster a customer-centric culture will thrive in this new landscape.

As you work toward this, you want to avoid the three pitfalls:

- Pitfall #1: Measuring the wrong thing: Organizations often focus on certain metrics, even if they are not aligned with the desired CX outcomes. AHT in contact centers can overshadow other important metrics.
- Pitfall #2: Measuring Frequently, Acting Infrequently: It is important to measure customer feedback at the right intervals as seen in the example of the company collecting customer surveys but failing to take action based on the feedback, resulting in dissatisfaction among customers.
- Pitfall #3: Doing what you've always done and expecting different results: You need to challenge your current practices and seek innovative approaches. Relying on traditional reports and metrics may not align with a customer-centric vision. So

it is important to evaluate the relevance and effectiveness of existing reports and metrics.

Consider whether you want to continue with the same practices or explore alternative approaches for better CX.

**Chapter 4: Data-Driven Insights and Rebuilding Culture**

This chapter sheds light on the power of data-driven insights and personalization in crafting remarkable experiences. We explore how leveraging customer data intelligently can enable organizations to anticipate needs, personalize interactions and build long-lasting relationships. We recognize the significance of robust data governance, ethics and security to build trust and respect customer privacy. In doing so, we emphasize that culture directly impacts CX and highlights the need for executive acknowledgment, support, and sponsorship for implementing widespread change. The chapter outlines the four steps you can take to prioritize CX:

1. Aligning to common goals
2. Identifying a responsible party
3. Establishing baselines on the seven key factors covered in Chapter 1
4. Removing barriers

Overall, the chapter emphasizes the significance of culture in driving customer-centricity and provides insights into the challenges and steps involved in building a customer-centric culture.

**Chapter 5: CX as a Strategic Differentiator**

In this final chapter, we reinforce that CX is no longer an optional addition but a strategic differentiator. We see how customer-centric organizations, driven by a relentless commitment to exceed customer expectations, outshine their competition. The chapter highlights the importance of evaluating the current state of the CX and identifying the desired future state. We suggest using workflows to map out customer interactions and understand their needs.

We lay out a three-step approach for implementing a CX transformation:

1. Focus on the Foundation — identifying the very basic steps to improve your customers' experience
2. Build the Bridge/Elevate the Experience — this step focuses on building the bridge to the desired end state
3. CX Nirvana — your desired end state including long-term customer relationships, residual revenue and a preoccupation with your customers' desires.

The chapter also introduces the implementation process based on the Deming Cycle (Plan-Do-Check-Act). We emphasize the need for alignment, planning, and communication within the organization. Visible work and cross-team collaboration are highlighted as essential for success.

## Embracing the future

As we reflect on the transformative journey we have undertaken, it becomes evident that the future of CX is brimming with possibilities. Organizations must continue to adapt, innovate and embrace change.

They must empower their employees, nurture a culture of continuous improvement and harness emerging technologies to deliver truly exceptional experiences.

You need to be a part of that change. We urge you to embark on your own CX transformation journey, leveraging the insights gained from this book as a guiding compass. With a relentless focus on customer-centricity, embracing digital disruption and nurturing the human touch, organizations can navigate the ever-changing landscape of CX and emerge as industry leaders.

We are aware everything is changing. In fact, at the time of this book's publication, we find ourselves in new, unchartered waters. We acknowledge generative AI has opened the door to revolutionizing the CX with its ability to understand and replicate human-like interactions. By leveraging generative AI technologies, businesses can provide personalized, efficient and empathetic customer support that surpasses traditional methods. Conversational AI models powered by generative algorithms can engage in natural, context-aware conversations with customers, answering inquiries, providing solutions and even handling complex issues. With generative AI, contact centers can scale their operations, improve response times and enhance customer satisfaction, leading to lasting positive impacts on the CX landscape.

The success of businesses in the future will hinge on their ability to place the customer at the center of everything they do. By embracing this new paradigm, organizations will not only thrive in an era defined by customer expectations but also create lasting, meaningful and profitable relationships with their customers.

# About the Authors

## Chad Hendren

Chad Hendren is a seasoned CX strategist and thought leader with a passion for driving business growth through exceptional customer interactions. With almost three decades of experience in the industry, Chad has worked with numerous global organizations, helping them transform their CX strategies to meet the evolving demands of the digital era.

Throughout his career, Chad has been at the forefront of the CX revolution, advocating for a customer-centric approach and leveraging emerging technologies to deliver personalized and memorable experiences. He is recognized for his expertise in digital transformation, data-driven insights and building customer-centric cultures within organizations.

As a prolific speaker, inventor and writer, Chad has shared his insights at industry conferences and contributed to leading publications. His expertise and thought leadership have inspired businesses worldwide to prioritize CX as a strategic differentiator. With his passion for driving positive change and his deep understanding of the evolving customer landscape, he is a trusted adviser in the field of CX.

Chad resides in McEwen, Tennessee, with his beautiful, supportive wife, Amy, and his mother.

## Gina Gibson

Gina Gibson is a dynamic and results-driven CX professional known for her ability to create transformative strategies that elevate brands and deliver exceptional customer journeys. With a strong background in project management, Gina brings a unique perspective to the world of CX, focusing on the intersection of emotional connection and business growth.

Throughout her career, Gina has worked with leading organizations across various industries, helping them harness the power of customer insights to drive innovation and deliver personalized experiences. She is highly skilled at recommending ideal customer outcomes and strategies to achieve and create meaningful interactions that build lasting relationships by leveraging data analytics and technology.

Gina is an adept workshop facilitator, sharing her expertise on CX strategy, journey mapping and customer-centric design. Her ability to blend analytical thinking with creative problem-solving enables organizations to reimagine their CXs and position themselves as industry leaders. With her passion for creating memorable experiences and her dedication to delivering tangible results, she is a trusted adviser for businesses seeking to excel in the realm of CX.

Gina resides in Scottsdale, Arizona.

Together, Chad Hendren and Gina Gibson bring a wealth of experience and expertise to the field of CX. Their collaborative approach and deep understanding of the evolving customer landscape have shaped their vision for a future where businesses can thrive by placing the customer at the heart of their strategies. Through their insights, guidance and thought leadership, Chad and Gina empower organizations to embrace the transformative power of CX and drive sustainable growth in an increasingly customer-centric world.

Printed in France by Amazon
Brétigny-sur-Orge, FR

18891459R00070